Vector Calculus

Formulations, Applications and Python Codes

Series in Computational Methods

Series Editor: Gui-Rong Liu *(University of Cincinnati, USA)*

Series in Computational Methods
Volume 4

Vector Calculus

Formulations, Applications and Python Codes

G. R. Liu

University of Cincinnati, USA

W|S World Scientific

NEW JERSEY · LONDON · SINGAPORE · BEIJING · SHANGHAI · HONG KONG · TAIPEI · CHENNAI · TOKYO

Published by

World Scientific Publishing Co. Pte. Ltd.

5 Toh Tuck Link, Singapore 596224

USA office: 27 Warren Street, Suite 401-402, Hackensack, NJ 07601

UK office: 57 Shelton Street, Covent Garden, London WC2H 9HE

Library of Congress Control Number: 2025010541

British Library Cataloguing-in-Publication Data
A catalogue record for this book is available from the British Library.

Series in Computational Methods — Vol. 4
VECTOR CALCULUS
Formulations, Applications and Python Codes

ISBN 978-981-98-1364-3 (hardcover)
ISBN 978-981-98-1365-0 (ebook for institutions)
ISBN 978-981-98-1366-7 (ebook for individuals)

For any available supplementary material, please visit
https://www.worldscientific.com/worldscibooks/10.1142/14328#t=suppl

Desk Editors: Nambirajan Karuppiah/Steven Patt

Typeset by Stallion Press
Email: enquiries@stallionpress.com

About the Author

Gui-Rong Liu received his Ph.D. from Tohoku University, Japan, in 1991. He was a Postdoctoral Fellow at Northwestern University, USA, from 1991 to 1993. He was a Professor at the National University of Singapore until 2010. He is currently a Professor at the Department of Aerospace Engineering and Engineering Mechanics, University of Cincinnati, USA. He was the Founder of the Association for Computational Mechanics (Singapore) (SACM) and served as the President of SACM until 2010. He served as the President of the Asia-Pacific Association for Computational Mechanics (APACM) (2010–2013) and an Executive Council Member of the International Association for Computational Mechanics (IACM) (2005–2010; 2020–2026). He authored a large number of journal papers and books, including two bestsellers, *Mesh Free Method: Moving Beyond the Finite Element Method* and *Smoothed Particle Hydrodynamics: A Meshfree Particle Methods*. He is the Editor-in-Chief of the *International Journal of Computational Methods* and served as an Associate Editor for *IPSE* and *MANO*. He is the recipient of numerous awards, including the Singapore Defence Technology Prize, NUS Outstanding University Researcher Award, NUS Best Teacher Award, APACM Computational Mechanics Award, JSME Computational Mechanics Award, ASME Ted Belytschko Applied Mechanics Award, Zienkiewicz Medal from APACM, AJCM Computational Mechanics Award, Humboldt Research Award, SACM Medal from the Association of Computational Mechanics (Singapore), and the Master Engineering Educator Award (UC, CEAS). He has been listed as one among the world's top 1% most influential scientists (Highly Cited Researchers) by Thomson Reuters for a number of years.

Contents

Chapter 1

Introduction

1.1 Computational methods

This book is part of a series exploring the exciting field of computational methods. Its goal is to provide a solid foundation in this discipline by combining theoretical principles with practical applications.

The series encompasses general methods and techniques used throughout science, technology, engineering, and mathematics (STEM) education, as well as research across various scientific and engineering disciplines. Consider it a comprehensive encyclopedia for computational methods, covering theory, formulation, and coding techniques necessary to get started. The series is designed to be accessible: Readers who have completed primary and lower secondary education (roughly nine years of schooling) can begin with the foundational volumes on mathematics. By progressing through the mechanics volumes, readers will be equipped to tackle research and design projects leveraging computational methods.

Taking advantage of rapid advancements in computer technology, code examples are heavily integrated throughout the series. This integration enables concepts, theories, and formulations to be vividly demonstrated through practical examples with clear visualizations. Initially, Python will be the primary language used in the foundational volumes, with the possibility of introducing other languages for more advanced topics. By effectively utilizing code, readers can dedicate more time to understanding core principles, as the computer manages complex calculations, tedious derivations of formulas, time-consuming operations, and extensive data processing. This approach allows the focus to remain on broader concepts.

This approach empowers readers to explore the fascinating world of computational methods with a strong theoretical grounding and practical skills.

1.2 Why start and contribute to this book series

The Editor-in-Chief of this book series, Dr. G.R. Liu, has been working in areas related to computational methods for over 40 years. He developed his first FEM code for nonlinear problems in 1980 and has since published more than 600 journal papers and over a dozen monographs in this area. After decades of studying, using, and developing computational methods, he began to envision a way to help others learn computational methods more effectively, systematically, and seamlessly. He concluded that developing this book series would be the best way to achieve this objective.

1.3 Vector calculus

Volume 1 of this book series covers two of the most important building blocks in computational methods: *Numbers and Functions: Theory, Application, and Python Codes* [2]. Volume 2 covers *Calculus: Formulation, Application, and Python Codes* [3], which provides powerful tools for differentiation and integration. This volume focuses on vector calculus, which offers essential tools to examine critical effects, extract significant features of a scalar or vector function, and reveal the mathematical descriptions of fundamental physical laws. This is achieved using various differential operations and integrations.

A vector function has both magnitude and direction at any point in its domain. It describes a field over the domain and is often called a **vector field**. To examine critical effects and extract key features, we use powerful tools of differentiation and integration. Depending on how the derivatives are performed, we can determine the **divergence**, **curl**, and **gradients** of a given field function, which reveal the mathematical behavior of the vector field. Through integration, applied in a properly designed manner while incorporating the geometric features of the integral domains, the effects of a vector field can be evaluated, including, for example, work done along a curve or surface and flux across a curve or surface. Integrals of these mathematical behaviors — divergence, curl, and gradient — over domains bounded by curves or surfaces can also be evaluated. Important relationships between these curve or surface integrals and the corresponding domain integrals were discovered by Green, Gauss, Stokes, and others, now known as Green's theorem, Stokes' theorem, and Gauss's formula. These foundational discoveries are gems of calculus and have laid the groundwork for modern computational methods. This book covers all these theorems and formulas in detail.

In essence, vector calculus employs differential operators (divergence, curl, and gradient) along with vector operators as 'stirring' tools and integral operators as summarizing tools for total effects. These theorems illustrate how these effects are interrelated.

Theories and concepts related to these theorems and formulas will be presented in detail. Techniques and Python codes for computing critical effects and examining these theorems and formulas will be introduced or developed for various types of vector functions and domains. Applications of these theorems and formulas will also be explored. Techniques and tools are developed for the examination and creation of special vector fields, such as potential fields, divergence-free fields, and fields that are both conservative and divergence-free. Techniques for approximating the gradient, curl, and divergence of functions without differentiation, known as gradient smoothing methods (GSMs), will also be presented, complementing the function approximation discussed in Volumes 1 and 2.

Python code is provided and used throughout the book, allowing readers to practice and interact with the material during their studies. This interactive approach makes it easier to comprehend the theories, concepts, formulations, and techniques associated with these theorems and formulas.

A strong connection between theory and real-world engineering problems is established through numerous examples.

The material in this book is suitable for various learning methods, including classroom teaching, online courses, and self-study. With the provided Python codes, readers can easily observe how theory is formulated and how solutions are achieved in terms of formulas, numerical results, and graphs. Readers can also deepen their understanding by experimenting with the codes and even developing their own solutions for related problems.

The book is written in Jupyter Notebook format, allowing seamless integration of theory, formulation, coding, and real-time interaction within a single document. This format provides an environment conducive to easy reading, exercises, practice, and further exploration.

This book draws upon works such as Refs. [4–7], which the author studied as a university student, while Ref. [8] serves as a reference book for self-study and research. Some example problems from these textbooks are included in this book. Both NumPy and SymPy are utilized in the development of code for the demonstration examples. Wikipedia has been a valuable source of reference, and interactions with ChatGPT, Gemini, and Bing have significantly contributed to the development of some codes and the preparation of this volume. The author's research experiences, insights, new perspectives, theories, formulations, algorithms, techniques, and utility codes are also integrated into this book. As a result, although vector calculus is a classic topic, it is presented here with novel discussions and modern tools, aiming to establish a solid foundation for effective computational methods in the analysis and design of advanced systems in science and engineering.

Numerous examples are presented in this book; however, the applications of vector calculus are vast, and the examples and case studies provided are not exhaustive.

In general, this book focuses on real functions unless specified otherwise.

1.4 Who may read this book

This book is written for beginners interested in learning computational methods for solving problems in science, engineering, and natural phenomena. The target audience includes high-school students, university students, graduate students, researchers, and professionals in any STEM discipline. Engineers and practitioners may also find the book useful for establishing a strong foundation in core computational methods and concepts.

For beginners, it is recommended to first read or skim through Volume 1 of this book series, *Numbers and Functions: Theory, Application, and Python Codes* [2], followed by Volume 2, *Calculus: Formulation, Application, and Python Codes* [3].

This book is written for human readers. It may not be used for training or testing any automated systems without the author's permission.

1.5 Codes used in this book

Readers who purchase the book may contact the author directly at liugr100@gmail.com to request a free soft copy of the book in Jupyter

Notebook format with codes (which may be updated) for academic use after registration.

Conditions of use: The following conditions apply to the use of the book and the codes developed by the author in hard copy, soft copy, and any other media and format:

- **User responsibility:** Users are solely responsible for any risks associated with the utilization of any part of the codes and techniques.
- **Code purpose:** The codes are primarily designed to demonstrate concepts and may not be optimized for efficiency or robustness. Many of the codes have not been tested by a third party.
- **Limited use:** The book and codes are for personal use only. Redistribution without the author's permission is not allowed.
- **No user support:** No support is provided for the codes.
- **Citation:** Proper citation and acknowledgment are required when using the book, codes, ideas, and techniques.

These codes often rely on various external packages and modules, and their behavior can be influenced by the versions of Python and these packages. If the code does not run as expected, a version mismatch could be the cause. To troubleshoot, check the versions of the packages and modules used. You can verify the versions within a code cell in the Jupyter Notebook. For example, to check the current Python environment version, you can use the following:

```
1  !python -V                            #! is used to execute an external command
```

Python 3.9.16

```
1  !jupyter notebook --version
```

6.1.5

1.6 Troubleshooting version issues

If you encounter a version mismatch error, you can either adapt the code to match the specific version or install the required version on your system. Searching the web using the error message may provide potential solutions or leads. Additionally, large language models can be helpful for troubleshooting.

This approach is the author's preferred method for resolving code execution issues.

1.7 Learning Python

This book does not cover Python basics, as many resources are available online. Interested readers can refer to Chapters 2 and 3 in Ref. [1] for a concise overview of using Python for scientific computations. Since Python is relatively easy to learn, you can utilize the codes and examples in this book to start learning Python while studying the technical subjects.

1.8 Use of external modules or dependencies

1.8.1 *Module imports in Jupyter notebooks*

Frequently used modules in a chapter are imported in the first code cell, which should be executed at the beginning. If you encounter a name error (such as `NameError: name 'sp' is not defined`), return to the beginning of the chapter and execute the cell, which typically reads as follows:

```
1   #Often used external modules are in commonImports placed in folder grbin
2   #Place cursor in this cell, and press Ctrl+Enter to import dependences.
3
4   import sys                          # for accessing the computer system
5   sys.path.append('../grbin/')            # Add in the path to your system
6
7   from commonImports import *        # Import dependences from '../grbin/'
8   import grcodes as gr                  # Import the module of the author
9   importlib.reload(gr)              # For in case, when grcodes is modified
10
11  from continuum_mechanics import vector
12  from continuum_mechanics.solids import sym_grad, strain_stress
13  init_printing(use_unicode=True)        # For latex-like quality printing
14
15  # Digits in print-outs
16  np.set_printoptions(precision=4,suppress=True,
17                      formatter={'float_kind': '{:.4e}'.format})
```

The above cell includes a common import file, `commonImports.py`, which contains all the frequently used modules in this book to avoid repeatedly importing lengthy external modules and reduce redundancy:

```
 1  # commonImports.py:
 2  from __future__ import print_function
 3  import numpy as np                      # for numerical computation
 4  import sympy as sp                       # sympy module for computation
 5  import numpy.linalg as lg               # numpy linear algebra module
 6  import scipy.linalg as sg               # scipy linear algebra module
 7  import scipy.integrate as si
 8  from scipy.stats import ortho_group
 9  import importlib
10  import itertools
11  import inspect
12  import csv
13  import pandas as pd
14  import random
15  from IPython.display import display, Math
16
17  import autograd.numpy as anp                      # Thinly-wrapped numpy
18  from autograd import grad
19
20  from grcodes import drawArrow, plotfig, printM, printx # frequently used
21
22  import math as ma
23  from sympy import sin, cos, tan, sinh, cosh, exp, log, symbols, lambdify
24  from sympy import pi, Matrix, sqrt, oo, integrate, diff,Derivative,latex
25  from sympy import MatrixSymbol,simplify,nsimplify,Function,init_printing
26  from sympy import factor, expand, nsimplify, Matrix, ordered, hessian
27  from sympy.plotting import plot as splt
28  init_printing(use_unicode=True) # for latex-like quality printing format
29
30  from matplotlib.ticker import MultipleLocator
31  import matplotlib.pyplot as plt          # for plotting figures
32  import matplotlib as mpl
33  from mpl_toolkits.mplot3d import Axes3D
34
35  #np.set_printoptions(formatter={'float': '{: 0.4f}'.format})
36  #np.set_printoptions(precision=8)#, suppress=True)
37  #cmap = {'complex_kind': '{:>6.3f}'.format}
38  #cmap = {'complex_kind': '{:13.3f}'.rstrip('0').rstrip('.').format}
39  #with np.printoptions(suppress=True,formatter=cMap):
40  #    print(f"Eigenvalues of skew-sy. part of random matrix A:\n{e}")
```

Module reloading: While importing the same module multiple times gen-erally doesn't cause issues, Python uses a cache and ignores subsequent imports if the module is already loaded. However, if you modify an imported module and want those changes to be reflected in your code, you need to reload the module. This can be done using the `importlib.reload()` function:

```
1  # grcodes was imported as gr, reload it when grcodes is modified
2  importlib.reload(gr)        # This is also included in the first code cell
```

To view the codes in the imported module, one may just use the following by uncommenting it (which may produce a long output):

```
1  import inspect
2  #source_code = inspect.getsource(gr)    # to view everything in gr module
3  source_code = inspect.getsource(gr.curl_vf) # to view any function in gr
4  print(source_code)
```

```
def curl_vf(vf, X):
    '''Compute the curl of a given vector function vf in Sympy w.r.t X.
       For 2D&3D.
    '''

    if len(vf) == 2:
        return sp.diff(vf[1], X[0]) - sp.diff(vf[0], X[1])

    curl_f = sp.Matrix([sp.diff(vf[2], X[1]) - sp.diff(vf[1], X[2]),
                        sp.diff(vf[0], X[2]) - sp.diff(vf[2], X[0]),
                        sp.diff(vf[1], X[0]) - sp.diff(vf[0], X[1])])
    return curl_f
```

Editing codes:

- You can use any text editor to view and modify the codes.
- Alternatively, copy the code into a cell within a Jupyter Notebook and edit it directly.

Using functions from imported modules: To use a function from an imported module, you typically combine the module name with the function name using a dot (.). For example, if you import a module named **gr** containing a function called **printx**, you would use it like this:

```
1  x = 8
2  gr.printx('x')          # when gr. is used, printx() is from grcodes module
```

```
x = 8
```

1.9 Use of help()

Exploring modules and objects: To access detailed information about a module or an object within a module after importing it, you can use the `help()` function. This is a useful way to explore the functionalities and attributes available in different modules. For example:

```
1  help(gr.gsm_polygon)
```

```
Help on function gsm_polygon in module grcodes:

gsm_polygon(f, X, nodes, p_out=True)
    Computes the smoothed gradient of a given scalar function f (Sympy)
    over a general polygonal domain defined by its nodes. Example:
    nodes = np.array([[-1,-1], [ 1,-1], [ 1, 1], [-1, 1]]).
    return: s_gradf - smoothed gradient of f.
            gradf - exact gradient of f.
```

Code comments: The code cells throughout the book include comments to provide additional explanations. These comments, denoted by a hash symbol (#) at the beginning of the line, are placed on the right side of the cell to minimize disruption to the code while offering quick reference and guidance when needed.

References

[1] G.R. Liu, *Machine Learning with Python: Theory and Applications*, World Scientific, New Jersey, 2023.
[2] G.R. Liu, *Numbers and Functions: Theory, Formulation, and Python Codes*, World Scientific, New Jersey, 2024.
[3] G.R. Liu, *Calculus: A Practical Course with Python*, World Scientific, New Jersey, 2025.
[4] Y.C. Fan, *Advanced Mathematics*, 1979.
[5] L.X. Yang and B.Y. Bi, *Analytic Mathematics Practices, by Boris Demidovich*, 2005.
[6] S. Timoshenko and J.N. Goodier, *Theory of Elasticity*, 1970. Available online http://books.google.com/books?id=yFISAAAAIAAJ&dq=theory+of+elasticity&ei=ICiKSsr3G4jwkQSbxMyPCg.
[7] Z.L. Xu, *Elasticity*, Vol. 1 & 2, People's Publisher, China, 1979.
[8] G. Strang, *Calculus*, Wellesley-Cambridge Press, Massachusetts, 1991.

Chapter 2

Preliminaries: Vectors and Operators

```
1  # Import necessary dependences.
2  import sys
3  sys.path.append('../grbin/')
4  from commonImports import *                    # import dependences
5  import grcodes as gr                           # import own modules
6  importlib.reload(gr)                      # when grcodes is modified
7  np.set_printoptions(precision=4,suppress=True,   # Digits in print-outs
8                      formatter={'float_kind': '{:.4e}'.format})
```

In Ref. [1], we discussed the differentiation and integration of scalar functions, where each function produces a single scalar value for given independent variables. This volume explores both the differentiation and integration of vector functions, where the function itself is a vector composed of multiple component functions. A vector function describes a 'field'; each component value and the direction of the vector change with the coordinates similar to how a scalar function varies. This forms a field of vectors over the domain, often referred to as a **vector field**.

Note that we are not merely treating each component function individually, as doing so would not differ from the discussions in Ref. [1]. Instead, we focus on cases where the component functions interact in meaningful ways, as seen in vector fields commonly encountered in science and engineering.

Operations on a vector function can take various forms. For instance, the multiplication of two vectors can involve the dot product, cross-product, elementwise product, or outer product. Depending on the type of derivative applied, we obtain **divergence**, **curl**, and **gradient**, each revealing different aspects of the vector function. In addition to calculating areas and volumes, we focus on the work done by vector fields and the flux resulting from them.

These effects typically depend on both path and direction. This chapter presents various operations on vector functions that will be used throughout this book.

This chapter references textbooks in Refs. [2–4]. Wikipedia has been a valuable source, and both NumPy and SymPy were utilized in developing code for the demonstration examples. Discussions with ChatGPT, Gemini, and Bing were also helpful in coding and preparing this chapter.

2.1 Representation of vectors

Since our objective is to discuss vector fields, it is most productive to first present preliminaries related to vectors and vector operations. We will not delve into extensive details but will list the major formulations and concepts used throughout this book. For more information, readers can refer to Chapter 2 in Ref. [4]. In this book, we assume that each component of a vector is generally a function. Even if a component is a constant, it is treated as a constant function. A vector has both magnitude and direction, and these properties vary within the domain.

In a general n-dimensional orthogonal coordinate space, a vector field $\mathbf{v}$ is expressed as

$$\mathbf{v} = \begin{bmatrix} v_1 \\ v_2 \\ \vdots \\ v_n \end{bmatrix} = v_1 \mathbf{i}_1 + v_2 \mathbf{i}_2 + \cdots + v_n \mathbf{i}_n \tag{2.1}$$

where v_i is the ith component of $\mathbf{v}$ and $\mathbf{i}_i$ is the **standard unit vector** along x_i-axis:

$$\mathbf{i}_1 = \begin{bmatrix} 1 \\ 0 \\ \vdots \\ 0 \end{bmatrix} ; \quad \mathbf{i}_2 = \begin{bmatrix} 0 \\ 1 \\ \vdots \\ 0 \end{bmatrix} ; \quad \cdots \quad \mathbf{i}_n = \begin{bmatrix} 0 \\ 0 \\ \vdots \\ 1 \end{bmatrix} \tag{2.2}$$

A vector field $\mathbf{v}$ in a three-dimensional (3D) coordinate space is often expressed more explicitly in Cartesian coordinates (x, y, z), as shown in Fig. 2.1(a).

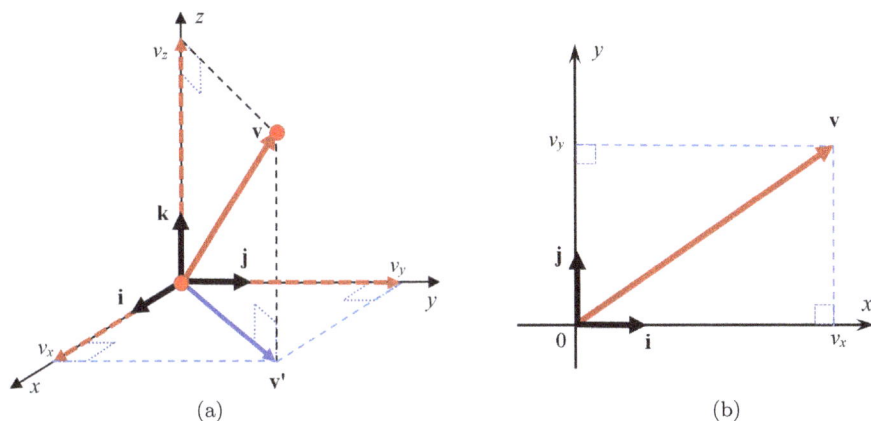

Figure 2.1. A vector in a coordinate space at the origin and its components generated by **orthogonal projections**: (a) 3D and (b) 2D.

In 3D cases, a vector field $\mathbf{v}$ is typically expressed as

$$\mathbf{v} = \begin{bmatrix} v_x \\ v_y \\ v_z \end{bmatrix} = v_x\mathbf{i} + v_y\mathbf{j} + v_z\mathbf{k} \tag{2.3}$$

where v_x, v_y, and v_z are the components of $\mathbf{v}$, obtained by projecting $\mathbf{v}$ orthogonally onto the x-, y-, and z-axes, respectively. The vectors $\mathbf{i}$, $\mathbf{j}$, and $\mathbf{k}$ are the three unit vectors along the x-, y-, and z-axes, respectively,

$$\mathbf{i} = \begin{bmatrix} 1 \\ 0 \\ 0 \end{bmatrix} ; \quad \mathbf{j} = \begin{bmatrix} 0 \\ 1 \\ 0 \end{bmatrix} ; \quad \mathbf{k} = \begin{bmatrix} 0 \\ 0 \\ 1 \end{bmatrix} \tag{2.4}$$

These are the basis vectors for a 3D space.

In a two-dimensional (2D) coordinate space, a vector $\mathbf{v}$ is shown in Fig. 2.1(b). It is expressed as

$$\mathbf{v} = \begin{bmatrix} v_x \\ v_y \end{bmatrix} = v_x\mathbf{i} + v_y\mathbf{j} \tag{2.5}$$

with the basis vectors given by

$$\mathbf{i} = \begin{bmatrix} 1 \\ 0 \end{bmatrix} ; \quad \mathbf{j} = \begin{bmatrix} 0 \\ 1 \end{bmatrix} \tag{2.6}$$

All these notation conventions are followed in this book.

2.2 Dot product of two vectors: A scalar

The dot product (or inner product) of two vectors $\mathbf{u}$ and $\mathbf{v}$ of the same length n is defined as

$$
\begin{aligned}
\mathbf{u} \cdot \mathbf{v} = \mathbf{v} \cdot \mathbf{u} = \mathbf{u}^\top \mathbf{v} &= \|\mathbf{u}\|\|\mathbf{v}\| \cos \theta \\
&= u_1 v_1 + u_2 v_2 + \cdots + u_n v_n
\end{aligned}
\tag{2.7}
$$

where θ is the angle between the two vectors. A distinct characteristic of the dot product is that it is the sum of the products of corresponding components of the two vectors, resulting in a scalar. The dot product is symmetric; thus, the order of the two vectors is immaterial.

> The dot product of two vectors results in a **scalar**. The dot product of two parallel vectors is the scalar product of the lengths of these two vectors. If they are in the same direction, the scalar product is positive; otherwise, it is negative.

> The dot product provides important information about how the two vectors are aligned or similar. It is a convenient measure of the **similarity** between two vectors.

The dot product is undoubtedly one of the most important operators and is widely used in computational methods. Many computer architectures are specifically designed for fast computation of dot products.

2.2.1 *Norm of a vector: Nonnegative, square root*

The norm of a vector is, in fact, induced from the dot product of the vector itself:

$$
\|\mathbf{v}\| = \sqrt{\mathbf{v} \cdot \mathbf{v}} = \sqrt{v_1 v_1 + v_2 v_2 + \cdots + v_n v_n} = \sqrt{\sum_{i=1}^{n} v_i^2}
\tag{2.8}
$$

Thus, it is always nonnegative. The norm is zero if and only if all components of $\mathbf{v}$ are zero. When the components of a vector are used to compute its norm, the introduction of a square root can sometimes give rise to difficulties in operations, such as in integration.

2.2.2 *Orthogonality of two vectors: Zero dot product*

The orthogonality of two vectors also arises from the dot product. Assuming $\mathbf{u}$ is orthogonal (perpendicular) to $\mathbf{v}$, the angle between them will be $\pm 90°$, leading to $\cos\theta = \cos(\pm 90°) = 0$. We thus have

$$\mathbf{u} \cdot \mathbf{v} = \|\mathbf{u}\|\|\mathbf{v}\| \times 0 = 0 \tag{2.9}$$

This implies the following:

> The dot product of two orthogonal vectors is zero, which is a useful property that is frequently made use of.

Let us take a look at some examples:

```
1  # Definition of variables and vectors in Sympy:
2
3  ux, uy, uz = sp.symbols('u_x, u_y, u_z', real=True) # define components
4  vx, vy, vz = sp.symbols('v_x, v_y, v_z', real=True)
5  u = sp.Matrix([ux, uy, uz])                    # define column vectors
6  v = sp.Matrix([vx, vy, vz])
7
8  display(Math(f" \\text{{Vector u = }}{latex(u)}"))
9  display(Math(f" \\text{{Dot-product of u and v = }}{latex(u.dot(v))}"))
10 display(Math(f" \\text{{Norm of vector u = }}{latex(u.norm())}"))
11 display(Math(f" \\text{{Norm of vector u =}}{latex(sqrt(sum(u.T@u)))}"))
```

Vector $\mathbf{u} = \begin{bmatrix} u_x \\ u_y \\ u_z \end{bmatrix}$

Dot-product of $\mathbf{u}$ and $\mathbf{v} = u_x v_x + u_y v_y + u_z v_z$

Norm of vector $\mathbf{u} = \sqrt{u_x^2 + u_y^2 + u_z^2}$

Norm of vector $\mathbf{u} = \sqrt{u_x^2 + u_y^2 + u_z^2}$

2.2.3 *Example: Mutually orthogonal vectors in 3D*

Consider the following vectors in 3D space. Examine their mutual orthogonality:

$$\mathbf{v_1} = \begin{bmatrix} 0 \\ -z \\ y \end{bmatrix} ; \quad \mathbf{v_2} = \begin{bmatrix} 0 \\ y \\ z \end{bmatrix} ; \quad \mathbf{v_3} = \begin{bmatrix} -y \\ 0 \\ x \end{bmatrix} ; \quad \mathbf{v_4} = \begin{bmatrix} x \\ 0 \\ y \end{bmatrix} \tag{2.10}$$

We use the following Python code to get it done:

```
1  x, y, z = symbols('x, y, z', real=True)  # define independent variables
2  v1 = sp.Matrix([0, -z, y])                        # define column vectors
3  v2 = sp.Matrix([0,  y, z])
4  v3 = sp.Matrix([-y, 0, x])
5  v4 = sp.Matrix([ x, 0, y])
6
7  display(Math(f" \\text{{$v_1 · v_2$ = }}{latex(v1.dot(v2))}, Yes."))
8  display(Math(f" \\text{{$v_3 · v_4$ = }}{latex(v3.dot(v4))}, Yes."))
9  display(Math(f" \\text{{$v_1 · v_3$ = }}{latex(v1.dot(v3))}, No."))
```

$\mathbf{v}_1 \cdot \mathbf{v}_2 = 0, \quad$ Yes

$\mathbf{v}_3 \cdot \mathbf{v}_4 = 0, \quad$ Yes

$\mathbf{v}_1 \cdot \mathbf{v}_3 = xy, \quad$ No

2.3 Cross-product of two vectors: A new vector

A cross-product of two vectors is defined as [4]

$$\mathbf{w} = \mathbf{u} \times \mathbf{v} = \underbrace{(u_y v_z - u_z v_y)}_{w_x}\mathbf{i} + \underbrace{(u_z v_x - u_x v_z)}_{w_y}\mathbf{j} + \underbrace{(u_x v_y - u_y v_x)}_{w_z}\mathbf{k}$$

$$= \begin{bmatrix} w_x \\ w_y \\ w_z \end{bmatrix} = \begin{bmatrix} u_y v_z - u_z v_y \\ u_z v_x - u_x v_z \\ u_x v_y - u_y v_x \end{bmatrix} \tag{2.11}$$

Note the following:

> The cross-product of two vectors produces a **vector**. The norm of this new vector represents the area of the parallelogram formed by the two vectors. The direction of the new vector is orthogonal to the plane formed by these two vectors, following the right-hand rule. Thus, we have $\mathbf{u} \times \mathbf{v} = -\mathbf{v} \times \mathbf{u}$.

If these two vectors $\mathbf{u}$ and $\mathbf{v}$ are in 2D, their cross-product becomes a scalar c:

$$c = \mathbf{u} \times \mathbf{v} = u_x v_y - u_y v_x \tag{2.12}$$

This scalar c corresponds to the third component of $\mathbf{u} \times \mathbf{v}$ in the 3D case.

A distinct characteristic of the cross-product is that each component of the result is a difference of the cross-products of the vector components. When we encounter this characteristic in an expression (often in this book), it likely arises from a cross-product operation.

To compute the cross-product of two vectors in 3D, we simply use the following:

```
1  display(Math(f" \\text{{uxv = }}{latex(u.cross(v))}"))
2  print(f" Is uxv=-vxu? {u.cross(v)== -v.cross(u)}")
```

$$\mathbf{u} \times \mathbf{v} = \begin{bmatrix} u_y v_z - u_z v_y \\ -u_x v_z + u_z v_x \\ u_x v_y - u_y v_x \end{bmatrix}$$

Is $\mathbf{u} \times \mathbf{u} = -\mathbf{v} \times \mathbf{u}$? True

One may also use gr.cross2v(), which accepts both 3D and 2D vectors:

```
1  display(Math(f" \\text{{uxv = }}{latex(gr.cross2v(u, v))}"))
2  print(f" Is uxv=-vxu? {gr.cross2v(u,v)== -gr.cross2v(v,u)}")
```

$$\mathbf{u} \times \mathbf{v} = \begin{bmatrix} u_y v_z - u_z v_y \\ -u_x v_z + u_z v_x \\ u_x v_y - u_y v_x \end{bmatrix}$$

Is $\mathbf{u} \times \mathbf{v} = -\mathbf{v} \times \mathbf{u}$? True

```
1  # Use slice to generate 2D vectors and compute their cross-product:
2  display(Math(f" \\text{{2D: uxv = }}{latex(gr.cross2v(u[:2], v[:2]))}"))
```

2D: $\mathbf{u} \times \mathbf{v} = u_x v_y - u_y v_x$

An inner product of two vectors produces a scalar, as seen in the previous section. In contrast, the outer product of two vectors produces a matrix. It has applications in linear algebra, particularly in matrix operations. In this book, we use it to generate the Jacobian matrix, which will be used for examining vector fields. The following is a general definition [4].

Let two vectors $\mathbf{u} \in \mathbb{R}^m$ and $\mathbf{v} \in \mathbb{R}^n$ be

$$\mathbf{u} = \begin{bmatrix} u_1 \\ u_2 \\ \vdots \\ u_m \end{bmatrix} ; \quad \mathbf{v} = \begin{bmatrix} v_1 \\ v_2 \\ \vdots \\ v_n \end{bmatrix} \tag{2.13}$$

The outer product $\mathbf{u} \otimes \mathbf{v}$ becomes a matrix in $\mathbb{R}^{m \times n}$:

$$
\underset{m \times n}{\mathbf{A}} = \underset{m \times 1}{\mathbf{u}} \otimes \underset{n \times 1}{\mathbf{v}}
$$

$$
= \begin{bmatrix} u_1 \\ u_2 \\ \vdots \\ u_m \end{bmatrix} \begin{bmatrix} v_1 & v_2 & \cdots & v_n \end{bmatrix} = \begin{bmatrix} u_1 v_1 & u_1 v_2 & \cdots & u_1 v_n \\ u_2 v_1 & u_2 v_2 & \cdots & u_2 v_n \\ \vdots & \vdots & \ddots & \vdots \\ u_m v_1 & u_m v_2 & \cdots & u_m v_n \end{bmatrix}
\tag{2.14}
$$

The (i,j)th element in $\mathbf{A}$ is $u_i v_j$. The outer product is equivalent to the matrix multiplication of $\mathbf{u}\mathbf{v}^\top$. Thus, the shapes of $\mathbf{u}$ and $\mathbf{v}$ are always compatible for this operation. This type of matrix is known as a **rank-one matrix** because it essentially has only one independent row or column.

Let us take a look at a few examples:

```
1  # Example: outer-product in 3D:
2  uOv =  u@v.T
3  display(Math(f" \\text{{Outer-product in 3D, u⊗v = }}{latex(uOv)}"))
```

$$
\text{Outer-product in 3D, } \mathbf{u} \otimes \mathbf{v} = \begin{bmatrix} u_x v_x & u_x v_y & u_x v_z \\ u_y v_x & u_y v_y & u_y v_z \\ u_z v_x & u_z v_y & u_z v_z \end{bmatrix}
$$

```
1  # Example: outer-product of a vector in 2D with one in 3D:
2  uOv =  u[:2,0]@v.T
3  display(Math(f" \\text{{Outer-product of u in 2D and v in 3D, u⊗v = }}\
4                                        {latex(uOv)}"))
```

$$
\text{Outer-product of } \mathbf{u} \text{ in 2D and } \mathbf{v} \text{ in 3D, } \mathbf{u} \otimes \mathbf{v} = \begin{bmatrix} u_x v_x & u_x v_y & u_x v_z \\ u_y v_x & u_y v_y & u_y v_z \end{bmatrix}
$$

2.4 Rotation matrix, changing directions

Considering a vector $\mathbf{v}$ in 2D is rotated by θ angle (positive counterclockwise), the new vector $\mathbf{v}'$ is computed using

$$
\mathbf{v}' = \mathbf{R}\mathbf{v}
\tag{2.15}
$$

where **R** is the rotation matrix given by

$$\mathbf{R} = \begin{bmatrix} \cos\theta & -\sin\theta \\ \sin\theta & \cos\theta \end{bmatrix} \tag{2.16}$$

If $\theta = \pm 90°$, $\mathbf{v}'$ will be orthogonal to $\mathbf{v}$. This is a quick way to make a vector that is orthogonal to a given vector. Following is an example:

```
1  def rotationR(θ):
2      '''Create a 2D sympy rotation matrix.
3      Input θ: rotation angle, symbolic. \
4      Return: Rotation matrix R of shape (2,2)'''
5      R = sp.Matrix([[ sp.cos(θ), -sp.sin(θ)],
6                     [ sp.sin(θ),  sp.cos(θ)]])
7      return R
```

```
1  def np_R(θ):
2      '''Create a 2D numpy rotation matrix.
3      Input θ: degree, rotation angle. \
4      Return: Rotation matrix R of shape (2,2)'''
5      R = np.array([[np.cos(np.deg2rad(θ)), -np.sin(np.deg2rad(θ))],
6                    [np.sin(np.deg2rad(θ)),  np.cos(np.deg2rad(θ))]])
7      return R
```

```
1  θ = symbols('θ', real=True)              # define an angle variable
2  vx, vy = symbols('v_x, v_y ', real=True)      # define components
3  v = sp.Matrix([vx, vy])                       # define column vectors
4
5  #R=Matrix([[cos(θ),-sin(θ)],[sin(θ),cos(θ)]])  # define rotation matrix
6
7  vp = gr.rotationR(θ)@v                        # use rotationR() directly
8  v90 = vp.subs(θ, sp.pi/2)
9  v90dot_v = (vp.dot(v)).subs(θ, sp.pi/2)             # dot-product
10 display(Math(f" \\text{{Vector v = }}{latex(v)}"))
11 display(Math(f" \\text{{v rotated by $90^o$ = }}{latex(v90)}"))
12 print(f" Is v90 orthogonal to v? {v90dot_v==0}")   # check orthogonality
```

Vector $\mathbf{v} = \begin{bmatrix} v_x \\ v_y \end{bmatrix}$

$\mathbf{v}$ rotated by $90° = \begin{bmatrix} -v_y \\ v_x \end{bmatrix}$

```
Is v90 orthogonal to v? True
```

As seen for vectors in 2D, when a vector is rotated by $(90°)$, its components swap along with a sign change to the first component. This is a useful fact for deriving formulas.

2.5 Gradient operator: A vector

The gradient operator ∇ is treated as a vector and has the following form:

$$\nabla \equiv \begin{bmatrix} \frac{\partial}{\partial x} \\ \frac{\partial}{\partial y} \\ \frac{\partial}{\partial z} \end{bmatrix} \tag{2.17}$$

Its components are differential operators (not numbers or functions). The gradient of a scalar function u becomes

$$\nabla u = \begin{bmatrix} \frac{\partial u}{\partial x} \\ \frac{\partial u}{\partial y} \\ \frac{\partial u}{\partial z} \end{bmatrix} \tag{2.18}$$

We note the following:

> The gradient of a scalar function becomes a vector (of functions).

2.6 Divergence of a vector: A scalar

The dot product $\nabla \cdot \mathbf{v}$ becomes a scalar function, called the divergence of $\mathbf{v}$, denoted as div; $\mathbf{v}$:

$$\text{div } \mathbf{v} \equiv \nabla \cdot \mathbf{v} = \frac{\partial v_x}{\partial x} + \frac{\partial v_y}{\partial y} + \frac{\partial v_z}{\partial z} \tag{2.19}$$

> The divergence of a vector, $\nabla \cdot \mathbf{v}$, becomes a scalar function. It quantifies how much the vector field is diverging (spreading out from or converging into a point) in the field. Thus, it relates to the magnitude of a source or sink in a vector field at a point for a given physical problem based on the physical conservation law.

For 2D cases, the divergence of $\mathbf{v}$ becomes

$$\text{div } \mathbf{v} \equiv \nabla \cdot \mathbf{v} = \frac{\partial v_x}{\partial x} + \frac{\partial v_y}{\partial y} \tag{2.20}$$

Since the divergence of a vector is a scalar at any point in the domain, it represents a scalar property of the vector at that point. Therefore, its value should not depend on the coordinate system used to describe it. This property is generally useful when performing a coordinate transformation in computational methods although we will not use it in this book.

2.7 Curl of a vector in 3D: A new vector

Consider a vector $\mathbf{v}$. The curl of $\mathbf{v}$ in a 3D space is defined using the cross-product of ∇ with $\mathbf{v}$:

$$\underbrace{\nabla \times \mathbf{v}}_{\text{curl } \mathbf{v}} = \begin{vmatrix} \mathbf{i} & \mathbf{j} & \mathbf{k} \\ \frac{\partial}{\partial x} & \frac{\partial}{\partial y} & \frac{\partial}{\partial z} \\ v_x & v_y & v_z \end{vmatrix} = \begin{bmatrix} \frac{\partial v_z}{\partial y} - \frac{\partial v_y}{\partial z} \\ \frac{\partial v_x}{\partial z} - \frac{\partial v_z}{\partial x} \\ \frac{\partial v_y}{\partial x} - \frac{\partial v_x}{\partial y} \end{bmatrix} \tag{2.21}$$

We note the following:

Curl; $\mathbf{v}$ is a vector. It measures the components of the **rate of circulation** at a point in the vector field with respect to the x-, y-, and z-axes, respectively. Its signature characteristic is that each component is a difference of cross-partial derivatives, resulting from the cross-product.

2.7.1 *Curl of a vector in 2D: Scalars*

Consider a vector $\mathbf{v}$ in a 2D space, specifically in the x–y plane. The curl of $\mathbf{v}$ is defined as

$$\text{curl}_z \mathbf{v} = \frac{\partial v_y}{\partial x} - \frac{\partial v_x}{\partial y} \tag{2.22}$$

This results in a scalar value, which is the z-component of the 3D curl, denoted as $\text{curl}_z \mathbf{v}$.

Similarly, for other 2D planes, we have

$$\text{curl}_y \mathbf{v} = \frac{\partial v_x}{\partial z} - \frac{\partial v_z}{\partial x}$$
$$\text{curl}_x \mathbf{v} = \frac{\partial v_z}{\partial y} - \frac{\partial v_y}{\partial z} \tag{2.23}$$

These too are scalars.

2.7.2 *Example: Physical meaning of a curl*

To better understand what a curl really is, let's consider a particle in the x–y plane that has an angular velocity of ω_z in the counterclockwise direction, as shown in Fig. 2.2. Based on simple geometry, we know that the particle at $(x, y, 0)$ should have a linear velocity of $v = r\omega_z$, and it should stay in the x–y plane.

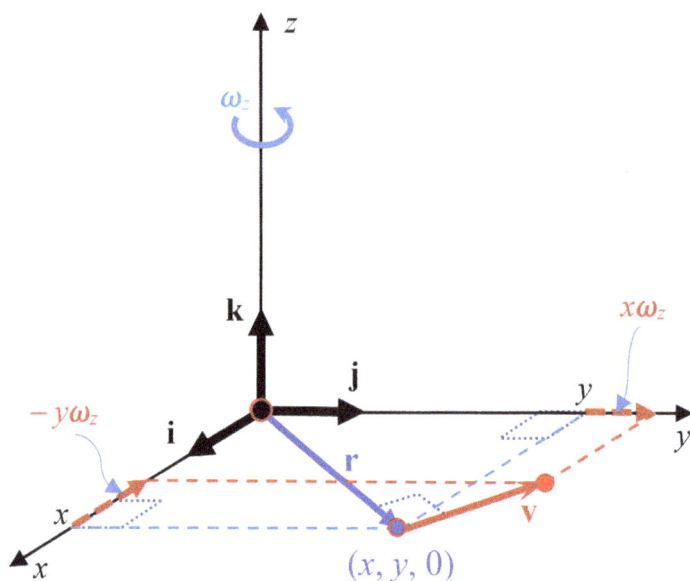

Figure 2.2.   Schematic view of a particle rotating about the z-axis.

To conduct a systematic analysis in 3D space for a particle, we can use vector calculations, which will lead to this result. First, define the position vector $\mathbf{r}$ of the particle:

$$\mathbf{r} = [x \ y \ 0]^\top \tag{2.24}$$

Next, define the vector for the angular velocity:

$$\omega = [0 \ 0 \ w_z]^\top \tag{2.25}$$

where w_z is the angular velocity about the z-axis.

Now, perform the cross-product between vectors ω and $\mathbf{r}$, which gives the velocity vector $\mathbf{v}$:

$$\mathbf{v} = \omega \times \mathbf{r} = [-yw_z \ xw_z \ 0]^\top \tag{2.26}$$

Let us take the curl of the velocity vector $\mathbf{v}$ using Eq. (2.21), then we obtain:

$$\text{curl } \mathbf{v} = \begin{bmatrix} \frac{\partial v_z}{\partial y} - \frac{\partial v_y}{\partial z} \\ \frac{\partial v_x}{\partial z} - \frac{\partial v_z}{\partial x} \\ \frac{\partial v_y}{\partial x} - \frac{\partial v_x}{\partial y} \end{bmatrix} = \begin{bmatrix} 0 \\ 0 \\ 2w_z \end{bmatrix} \tag{2.27}$$

It is clear that the curl of the velocity vector $\mathbf{v}$ of a particle in the x–y plane that rotates about the z-axis is twice its angular velocity. It captures the angular or rotational characteristics of the velocity vector of the particle. The Python snippet to achieve this is as follows:

```
1  x, y, z = symbols('x, y, z', real=True)          # define variables
2  X = sp.Matrix([x, y, z])                          # coordinate vector
3  r = sp.Matrix([x, y, z])
4
5  wz = symbols('w_z', real=True)                    # angular velocity about z-axis
6  w  = sp.Matrix([0, 0, wz])                        # vector function
7
8  v = w.cross(r)                                    # cross-product
9  curl_v = gr.curl_vf(v, X)              # The curl of a vector function
10
11 display(Math(f" \\text{{Velocity v =}}{latex(v)};\;\;\;\
12                \\text{{Curl of v =}}{latex(curl_v)}"))
```

$$\text{Velocity } \mathbf{v} = \begin{bmatrix} -y\omega_z \\ x\omega_z \\ 0 \end{bmatrix} ; \quad \text{Curl of } \mathbf{v} = \begin{bmatrix} 0 \\ 0 \\ 2\omega_z \end{bmatrix}$$

If we compute the divergence of the velocity vector of the rotating particle, we shall have the following:

```
1  div_v  = gr.div_vf(v, X)              # Divergence of a vector function
2  display(Math(f" \\text{{Divergence of v =}}{latex(div_v)}"))
```

Divergence of $\mathbf{v} = 0$

We got zero. This is because the particle is rotating about the z-axis and will not diverge away from the origin.

Readers may take time to study this example so that the concepts of cross-product, curl, and divergence can make good sense. They are encouraged to conduct the same analysis for a particle rotating about the x-axis, paying attention to direction issues.

With these preliminary concepts and formulas, we should be able to comfortably study vector fields. The only care needed is that each component of a vector is now a function of coordinates and is hence subject to differentiation and integration.

Let us take a look at some examples involving various operators.

2.8 Example: Gradient, divergence, and curl operators

The following code demonstrates all these operations:

```
1  x, y, z = symbols('x, y, z', real=True)            # define variables
2  X = sp.Matrix([x, y, z])                           # coordinate vector
3
4  u = Function('u', real=True)(x, y, z)              # scalar function
5
6  vx = Function('v_x', real=True)(x, y, z)
7  vy = Function('v_y', real=True)(x, y, z)
8  vz = Function('v_z', real=True)(x, y, z)
9  v = sp.Matrix([vx, vy, vz])                        # vector function
10
11 grad_u = gr.grad_f(u, X)              # Gradient of a scalar function
12 div_v  = gr.div_vf(v, X)              # Divergence of a vector function
13 curl_v = gr.curl_vf(v, X)            # The curl of a vector function
14
15 display(Math(f" \\text{{Gradient of u =}}{latex(grad_u)}"))
16 display(Math(f" \\text{{Divergence of v =}}{latex(div_v)}"))
17 display(Math(f" \\text{{Curl of v =}}{latex(curl_v)}"))
```

$$\text{Gradient of u} = \begin{bmatrix} \frac{\partial}{\partial x}u(x,y,z) \\ \frac{\partial}{\partial y}u(x,y,z) \\ \frac{\partial}{\partial z}u(x,y,z) \end{bmatrix}$$

$$\text{Divergence of } \mathbf{v} = \frac{\partial}{\partial x}v_x(x,y,z) + \frac{\partial}{\partial y}v_y(x,y,z) + \frac{\partial}{\partial z}v_z(x,y,z)$$

$$\text{Curl of } \mathbf{v} = \begin{bmatrix} -\frac{\partial}{\partial z}v_y(x,y,z) + \frac{\partial}{\partial y}v_z(x,y,z) \\ \frac{\partial}{\partial z}v_x(x,y,z) - \frac{\partial}{\partial x}v_z(x,y,z) \\ -\frac{\partial}{\partial y}v_x(x,y,z) + \frac{\partial}{\partial x}v_y(x,y,z) \end{bmatrix}$$

2.9 Example: Curl of the gradient of a scalar function

As mentioned earlier, the gradient of a scalar function is a vector. It is thus interesting to take a look at the curl of the gradient of a scalar function:

```
1  curl_grad_u = gr.curl_vf(grad_u, X)    # curl of gradient of a function
2  curl_grad_u
```

$$\begin{bmatrix} 0 \\ 0 \\ 0 \end{bmatrix}$$

We obtain the widely used identity of vector calculus:

$$\boxed{\nabla \times \nabla u = \mathbf{0}} \tag{2.28}$$

> The curl of the gradient of any scalar function is a zero vector. In other words, a **gradient field**, which is a special field with components from the derivatives of a scalar function, is curl-free.

Due to this, a gradient field is also said to be irrotational.

2.10 Example: Laplace operator — Divergence of gradient

```
1  gr.div_vf(grad_u, X)          # divergence of the gradient of a function
```

$$\frac{\partial^2}{\partial x^2}u(x,y,z) + \frac{\partial^2}{\partial y^2}u(x,y,z) + \frac{\partial^2}{\partial z^2}u(x,y,z)$$

This is the famous Laplace operator $\nabla^2 \equiv \nabla \cdot \nabla = \frac{\partial^2}{\partial x^2} + \frac{\partial^2}{\partial y^2} + \frac{\partial^2}{\partial z^2}$ on a scalar function, which relates the source term of many physical problems, such as heat sources, to a temperature field (Chapter 14 in Ref. [5]), body forces of a stress field [4], and electric charge density to an electric field.

2.11 Example: Divergence of the curl of a vector function

As mentioned earlier, the curl of a vector function is a vector. Thus, it is interesting to see the divergence of the curl:

```
1  curl_v = gr.curl_vf(v, X)             # The curl of the vector function
2                                         # should produce a vector function
3  gr.div_vf(curl_v, X)   # The divergence of the curl of a vector function
4                                         # should produce a scalar
```

0

We obtain another widely used identity of calculus:

$$\boxed{\nabla \cdot \nabla \times \mathbf{v} = 0} \tag{2.29}$$

> The divergence of the curl of any vector function is zero. In other words, a solenoidal field or spin field, which is a special field with components from the curl of a vector function, is divergence-free.

A solenoidal field is also said to be incompressible because of the zero divergence in any direction, implying that there is no change in volume.

These identities are useful in derivations among scalar and vector functions.

2.12 Gradient of a vector function

The divergence of a vector function produces a scalar, and the curl of a vector function produces a vector. What do we get if a gradient operator is applied to a vector? The answer is a matrix.

2.12.1 *Definition*

Consider a general differentiable vector field in 3D Cartesian coordinates defined by a vector of three scalar functions:

$$\mathbf{v}(\mathbf{x}) = v_x(\mathbf{x})\mathbf{i} + v_y(\mathbf{x})\mathbf{j} + v_z(\mathbf{x})\mathbf{k} = \begin{bmatrix} v_x(\mathbf{x}) \\ v_y(\mathbf{x}) \\ v_z(\mathbf{x}) \end{bmatrix} \tag{2.30}$$

where $v_i(\mathbf{x})$ $(i = x, y, z)$ are the components of the vector function $\mathbf{v}(\mathbf{x})$. The vector function creates a vector field in the domain where these functions are defined. Each of the component functions is a function of the coordinates and can be subjected to partial differentiations. The gradient of a vector function is thus defined using the **outer product** of ∇ and $\mathbf{v}(\mathbf{x})$ so that all these component functions can be partially differentiated:

$$\begin{aligned}
\nabla\mathbf{v}(\mathbf{x}) &:= \underbrace{\begin{bmatrix} \frac{\partial}{\partial x} \\ \frac{\partial}{\partial y} \\ \frac{\partial}{\partial z} \end{bmatrix}}_{\nabla} \otimes \underbrace{\begin{bmatrix} v_x(\mathbf{x}) \\ v_y(\mathbf{x}) \\ v_z(\mathbf{x}) \end{bmatrix}}_{\mathbf{v}(\mathbf{x})} \\[2em]
&= \begin{bmatrix} \frac{\partial}{\partial x} \\ \frac{\partial}{\partial y} \\ \frac{\partial}{\partial z} \end{bmatrix} \begin{bmatrix} v_x(\mathbf{x}) & v_y(\mathbf{x}) & v_z(\mathbf{x}) \end{bmatrix} \\[2em]
&= \begin{bmatrix} \frac{\partial}{\partial x}v_x(\mathbf{x}) & \frac{\partial}{\partial x}v_y(\mathbf{x}) & \frac{\partial}{\partial x}v_z(\mathbf{x}) \\ \frac{\partial}{\partial y}v_x(\mathbf{x}) & \frac{\partial}{\partial y}v_y(\mathbf{x}) & \frac{\partial}{\partial y}v_z(\mathbf{x}) \\ \frac{\partial}{\partial z}v_x(\mathbf{x}) & \frac{\partial}{\partial z}v_y(\mathbf{x}) & \frac{\partial}{\partial z}v_z(\mathbf{x}) \end{bmatrix}
\end{aligned} \tag{2.31}$$

This leads to a matrix (of functions). For a given vector function, it can be generated using the following:

```
1  g_vf = gr.grad_vf(v, X)
2  display(Math(f" \\text{{Gradient of vector function v}}={latex(g_vf)}"))
```

$$\text{Gradient of vector function } \mathbf{v} = \begin{bmatrix} \frac{\partial}{\partial x}v_x(x,y,z) & \frac{\partial}{\partial x}v_y(x,y,z) & \frac{\partial}{\partial x}v_z(x,y,z) \\ \frac{\partial}{\partial y}v_x(x,y,z) & \frac{\partial}{\partial y}v_y(x,y,z) & \frac{\partial}{\partial y}v_z(x,y,z) \\ \frac{\partial}{\partial z}v_x(x,y,z) & \frac{\partial}{\partial z}v_y(x,y,z) & \frac{\partial}{\partial z}v_z(x,y,z) \end{bmatrix}$$

2.12.2 *Python examples*

Let us evaluate the gradient of a vector function, assuming each of the component functions is differentiable, such as the vector functions given in the following:

```
1  vf = Matrix([y*sp.sin(x), z*sp.sin(y), x*sp.cos(z)])
2  display(Math(f" \\text{{Example vector function = }}{latex(vf)}"))
```

$$\text{Example vector function} = \begin{bmatrix} y\sin(x) \\ z\sin(y) \\ x\cos(z) \end{bmatrix}$$

```
1  # The gradient of the vector function:
2  g_vf = gr.grad_vf(vf, X)      # produces 3 column vectors of 3 functions
3  display(Math(f" \\text{{Gradient of a vector function=}}{latex(g_vf)}"))
```

$$\text{Gradient of a vector function} = \begin{bmatrix} y\cos(x) & 0 & \cos(z) \\ \sin(x) & z\cos(y) & 0 \\ 0 & \sin(y) & -x\sin(z) \end{bmatrix}$$

It is a 3×3 matrix. This is because we have three component functions, and the gradient of each of the functions is a vector. These three vectors form a matrix.

2.12.3 *Jacobian matrix*

Computation with matrices is a huge topic and deserves separate volumes. Here, we just introduce one of the matrices that will be used frequently in this book: the **Jacobian matrix**. It is one of the most frequently used matrices in computational methods.

For a vector field $\mathbf{v}(\mathbf{x}) = [v_x(x, y, z); v_y(x, y, z); v_z(x, y, z)]^\top$, the Jacobian matrix $\mathbf{J}$ at a point $\mathbf{x}_0 = (x_0, y_0, z_0)$ is defined as

$$
\mathbf{J}(\mathbf{x}_0) = \begin{vmatrix} \frac{\partial v_x}{\partial x} & \frac{\partial v_x}{\partial y} & \frac{\partial v_x}{\partial z} \\ \frac{\partial v_y}{\partial x} & \frac{\partial v_y}{\partial y} & \frac{\partial v_y}{\partial z} \\ \frac{\partial v_z}{\partial x} & \frac{\partial v_z}{\partial y} & \frac{\partial v_z}{\partial z} \end{vmatrix}_{\mathbf{x}_0}
\tag{2.32}
$$

Clearly, the Jacobian matrix is a transpose of the gradient of the vector function. Thus, we can generate it using `gr.grad_vf()` for a given vector field. We shall do so frequently in later chapters.

The Jacobian matrix contains partial derivatives of each of the component functions of a vector field. Therefore, it collectively provides insight into how the vector field changes near that point. For a given vector function, Eq. (2.32) can be formed and then used to compute the eigenvalues of $\mathbf{J}$ at a point of interest $\mathbf{x}_0$. These eigenvalues reveal the following property of the vector field at the point:

- If the eigenvalues are real and positive, the vector field flows away from the point. It is repulsive.
- If they are real and negative, the vector field flows into the point. It is attractive.
- If the eigenvalues are real and have opposite signs, the vector field flows into the point in some directions and out from it in other directions. The point is a saddle point [1].
- If the eigenvalues are complex-valued at a point in the field, the point is a spiral point.
- If all eigenvalues are zero, this means that the point is neither a saddle, attractive, nor repulsive or spiral. The behavior of the vector field near this point cannot be determined from the Jacobian. A nonlinear analysis may be needed.

For 2D vector fields, the Jacobian matrix $\mathbf{J}$ at a point $\mathbf{x}_0 = (x_0, y_0)$ becomes

$$
\mathbf{J}(\mathbf{x}_0) = \begin{vmatrix} \frac{\partial v_x}{\partial x} & \frac{\partial v_x}{\partial y} \\ \frac{\partial v_y}{\partial x} & \frac{\partial v_y}{\partial y} \end{vmatrix}_{\mathbf{x}_0}
\tag{2.33}
$$

The Jacobian matrix will be used frequently in the following chapters for examining vector fields. Note that for the foregoing analysis, one can simply use the matrix of the gradient of the vector function because the transpose of a matrix does not affect the eigenvalues of the matrix.

2.13 Concluding remarks

This chapter briefs some of the most basic operators on vectors, which will be used in this book. It is worth being aware that a dot product produces a scalar, a cross-product produces a vector, and an outer product produces a matrix. This simple fact shows that mathematics can be interesting in addition to being useful.

Another product of vectors is the so-called **elementwise product**, implying that each element in two vectors is multiplied, respectively. It produces a vector. The sum of the elements in the vector gives the dot product. It is one of the standard operations in Python and is frequently used in computations, including machine learning [6]. It is more often used in NumPy.

Dot product, **cross-product**, **outer product**, **divergence**, **curl**, **gradient**, and the **Jacobian matrix** are the basic tools we use for operating and examining vector fields.

The following chapter discusses curve integrals of vector functions and presents ways in which a vector function can be turned into a scalar function and integrated along a given curve, producing, finally, just a scalar number.

References

[1] G.R. Liu, *Calculus: A Practical Course with Python*, World Scientific, New Jersey, 2025.
[2] G. Strang, *Calculus*, Wellesley-Cambridge Press, Massachusetts, 1991.
[3] G.R. Liu, *Numbers and Functions: Theory, Formulation, and Python Codes*, World Scientific, New Jersey, 2024.
[4] G.R. Liu, *Mechanics of Materials: Formulations and Solutions with Python*, World Scientific, New Jersey, 2024.
[5] G.R. Liu and T.T. Nguyen, *Smoothed Finite Element Methods*, Taylor and Francis Group, New York, 2010.
[6] G.R. Liu, *Machine Learning with Python: Theory and Applications*, World Scientific, New Jersey, 2023.

Chapter 3

Integration of Vector Fields along Curves

```
1  # Place cursor in this cell, and press Ctrl+Enter to import dependences.
2  import sys                           # For accessing the computer system
3  sys.path.append('../grbin/')   # Add in the path to your system
4
5  from commonImports import *        # Import dependences from '../grbin/'
6  import grcodes as gr                 # Import the module of the author
7  importlib.reload(gr)                 # When grcodes is modified, reload it
8
9  init_printing(use_unicode=True)      # For latex-like quality printing
10 np.set_printoptions(precision=4,suppress=True,
11                     formatter={'float_kind': '{:.4e}'.format})
```

This chapter first introduces some typical vector fields that will be used frequently in this book. Techniques and Python code to examine and extract features of these fields via integration along curves will be presented. We consider two important cases: the effect of a vector field along a curve, called **work** done, and that across a curve defined in the domain, called **flux**. Both cases have significance in applications in science and engineering.

This chapter is written with reference to textbooks in Refs. [1–3]. Wikipedia pages, particularly those on integral (https://en.wikipedia.org/wiki/Integral), serve as valuable additional references.

Both NumPy and SymPy are used in the development of code for the demonstration examples. Discussions with ChatGPT, Gemini, and Bing have also greatly helped in coding and in the preparation of this chapter.

3.1 Vector functions

Consider a vector field in 3D Cartesian coordinate space defined by a vector of three scalar functions:

$$\mathbf{f}(\mathbf{x}) = f_x(\mathbf{x})\mathbf{i} + f_y(\mathbf{x})\mathbf{j} + f_z(\mathbf{x})\mathbf{k} = \begin{bmatrix} f_x(\mathbf{x}) \\ f_y(\mathbf{x}) \\ f_z(\mathbf{x}) \end{bmatrix} \tag{3.1}$$

where $\mathbf{i}$, $\mathbf{j}$, and $\mathbf{k}$ are, respectively, the unit basis vectors on the x-, y-, and z-axes and $f_i(\mathbf{x})$; $(i = x, y, z)$ are the components of the vector function $\mathbf{f}(\mathbf{x})$ with $\mathbf{x} = (x, y, z)$. For a 2D problem, we simply drop everything related to the z-axis.

All vectors and matrices are written in boldface, including the vector of coordinates. When a vector field is said to have a certain property (such as differentiability or integrability), it implies that each and all of the component functions have the property. The vector function creates a vector field in the domain $\mathcal{D}$ where all its component functions are defined. Each of the component functions is, in general, a function of multiple coordinates.

Here, we use the generic name $\mathbf{f}$ for a vector field. Based on the features of a vector field, we may name it differently. For example, for a force field, we may use $\mathbf{F}$; for a velocity field, we may use $\mathbf{V}$, and so on.

Let us introduce some useful vector fields, starting with the position field denoted as $\mathbf{r}$.

3.1.1 *Position vector field*

The position vector field in the domain $\mathcal{D}$, defined in Cartesian coordinates, is expressed as

$$\mathbf{r}(x, y) = \begin{bmatrix} r_x \\ r_y \end{bmatrix} = x\mathbf{i} + y\mathbf{j} = \begin{bmatrix} x \\ y \end{bmatrix} \quad \text{for 2D}$$

$$\mathbf{r}(x, y, z) = \begin{bmatrix} r_x \\ r_y \\ r_z \end{bmatrix} = x\mathbf{i} + y\mathbf{j} + z\mathbf{k} = \begin{bmatrix} x \\ y \\ z \end{bmatrix} \quad \text{for 3D} \tag{3.2}$$

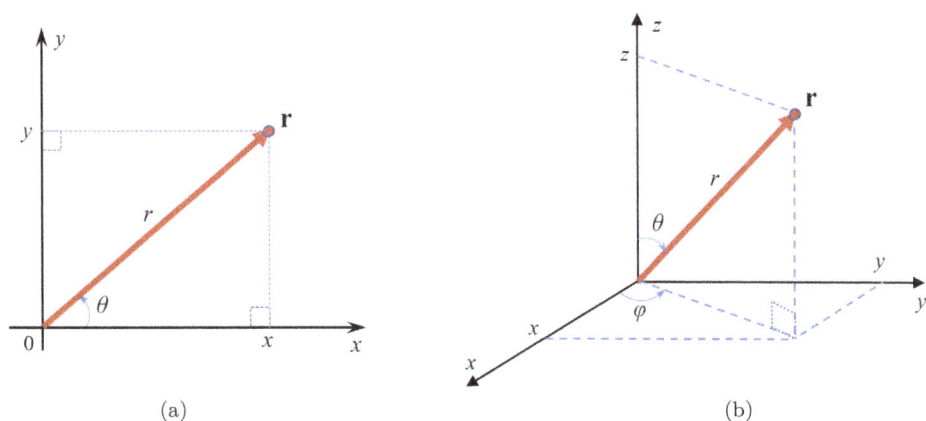

Figure 3.1. Position vectors defined in (a) 2D and (b) 3D Cartesian coordinates with the centers of radiation set at the origin. The position vectors can also be expressed in polar (2D) and spherical (3D) coordinate systems.

These vectors are shown in Fig. 3.1. In 3D cases, the component functions are very simple: $r_x(x, y, z) = x$, $r_y(x, y, z) = y$, and $r_z(x, y, z) = z$. The position vector field defines the position of a point (x, y, z) in 3D space, as shown in Fig. 3.1. For 2D cases, we simply drop everything related to z.

The vector $\mathbf{r}$ is also simply called the **position field**. When the reference point of $\mathbf{r}$ is set at the origin $(0, 0, 0)$ of the Cartesian coordinate system, the position vector $\mathbf{r}$ is the same as the coordinate vector $\mathbf{x}$. The length of the position vector with its tip at (x, y, z) is its norm:

$$|\mathbf{r}(x, y, z)| = \sqrt{x^2 + y^2 + z^2} = r \tag{3.3}$$

Note that $\mathbf{r}$ is a vector and r is a scalar (radial coordinate). Since the position vector radiates outward from its center, it is also a **radial field**. Note that the reference point (or the center of radiation) for a position vector does not have to be at $(0, 0, 0)$. We often choose a reference point that is convenient for operations.

3.1.2 *Python code to plot vector fields*

Let us write a Python function to plot the vector fields, which will be used multiple times (readers may skip reading it for now and return to it later if they encounter issues using it):

```
 1  def plotField(a, b, fx, fy, *p_name, scale=10):
 2      '''
 3          Plot a vector field with given two component numpy functions
 4          f=(fx, fy) in domain [-a, a]×[-b, b]
 5          p_name: filename in str when saving the plot to a file.
 6          scale: for adjusting the size of arrows.
 7      '''
 8      dx = a/10.; dy = b/10.
 9
10      X = np.arange(-a, a+dx, dx)
11      Y = np.arange(-b, b+dy, dy)
12      X, Y = np.meshgrid(X, Y)
13
14      plt.figure(); plt.ioff()
15      mag = np.sqrt(X**2 + Y**2)                            # magnitude
16      plt.quiver(X, Y, fx(X,Y), fy(X,Y), mag, scale=scale, cmap=cm.jet)
17
18      # Draw a circular curve in the field:
19      θ = np.linspace(0, 2*np.pi, 200)
20      r_circle = 0.9*a
21      plt.plot(r_circle*np.cos(θ), r_circle*np.sin(θ))
22
23      # Arrows on curve, normal n and tangent t directions:
24      α = 0.3; n = np.pi/6
25      nx_a = r_circle*np.cos(n); ny_a = r_circle*np.sin(n)
26      nx_b = α*nx_a;               ny_b = α*ny_a
27      plt.arrow(nx_a, ny_a, nx_b, ny_b, width=0.004, lw=2, color='r')
28      plt.text(nx_a+nx_b,.8*(ny_a+ny_b),r'$\mathbf{n}$',c='r',fontsize=10)
29
30      tx_b, ty_b = gr.np_R(90)@np.array([nx_b, ny_b]) # rotate n to get t
31      plt.arrow(nx_a, ny_a, tx_b, ty_b, width=0.004, lw=2, color='r')
32      plt.text(1.1*(nx_a+tx_b),ny_a+ty_b,r'$\mathbf{t}$',c='r',fontsize=10)
33
34      plt.axis('equal')  #plt.colorbar()
35      plt.savefig('imagesVC/'+p_name[0]+'.png', dpi=500)   # save to file
36      return plt
```

Let us plot the 2D position field defined in Eq. (3.2):

```
 1  from matplotlib import cm                    # needed for the plot
 2  plt.rcParams.update({'font.size': 8})
 3  plt.rcParams['figure.dpi'] = 500
 4  a, b = 1., 1.                                # domain [-a, a]×[-b, b] for plots
 5
 6  fx = lambda x, y: x + y*0    # the position field as functions of x and y
 7  fy = lambda x, y: x*0 + y    # use 0, so that f has both x,y variables
 8
 9  plt = plotField(a, b, fx, fy, 'positionField')  # plot file name as str
10  # fig, ax = plt.gca().figure, plt.gca()    # for modification if needed
11  #plt.show()
```

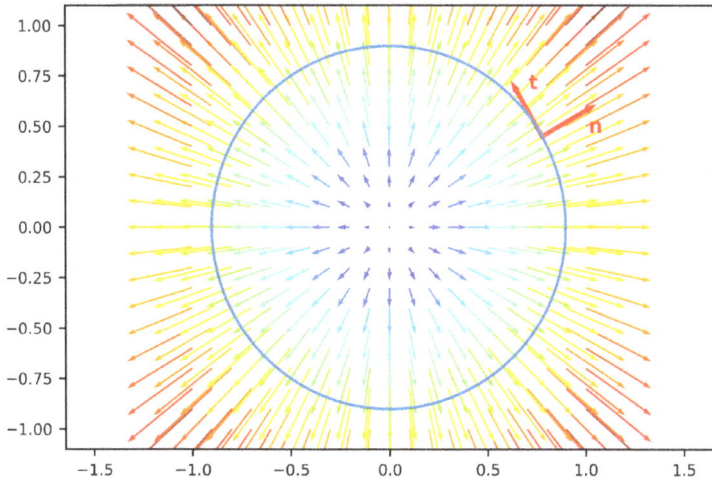

Figure 3.2. The position vector field radiating outward from its center.

The most significant feature of a position vector field is that it radiates outward from its center, as shown in Fig. 3.2. Hence, it has no circulation in the entire domain.

We have also plotted a circular curve in the field. The unit normal vector **n** and the tangent vector **t** at a point on the curve are also shown. It can be seen that the field vectors are in the same direction as **n**. The component of the field vectors in the tangent direction of the circle is zero.

Let us compute the curl of the position field along with its divergence:

```
1  x, y = symbols('x, y ', real=True)              # define variables
2  X = sp.Matrix([x, y])                       # the 2D cooridnate vector
3  r = sp.Matrix([x, y])                       # a 2D position vector
4
5  display(Math(f" \\text{{curl of the position vector = }}\
6                      {latex(gr.curl_vf(r, X))}"))
7  display(Math(f" \\text{{divergence of the position vector = }}\
8                      {latex(gr.div_vf(r, X))}"))
```

Curl of the position vector $= 0$

Divergence of the position vector $= 2$

It is found that the curl of the position vector is zero. This is expected because the position (radial) vector does not have any circulation in the tangent direction of any circle centered at $(0, 0)$.

On the other hand, the divergence of the position vector is a constant 2. This implies that the position vector field is diverging (spreading out) from any point in the field in the radial direction at a rate of 2. This also means that if this field exists, there must be sources in the field at every point to ensure that the physical conservation law holds.

Let us now check its Jacobian matrix defined in Eq. (2.33). We will use the gradient to generate the Jacobian matrix (with a transpose):

```
1  #J=sp.Matrix([[r[0].diff(x),r[0].diff(y)],[r[1].diff(x),r[1].diff(y)]])
2  J = gr.grad_vf(r, X).T    # use gradient to generate the Jacobian matrix
3  display(Math(f" \\text{{Jacobian matrix of r = }}{latex(J)}"))
```

Jacobian matrix of $\mathbf{r} = \begin{bmatrix} 1 & 0 \\ 0 & 1 \end{bmatrix}$

This matrix is very simple, with its two eigenvalues located on the diagonal. They are both positive at any point in the entire field, indicating that this vector field is repulsive (see Section 2.12.3). If the vector field represents fluid flow, the fluid at any point flows away from that point. If the vector field represents a force field, a particle at any point is pushed away from that point.

If the sign of the position field is reversed,

$$\mathbf{r}(x, y) = \begin{bmatrix} -x \\ -y \end{bmatrix} \tag{3.4}$$

We then have the following:

```
1  r = sp.Matrix([-x,-y])                    # a reversed radial field
2  J = gr.grad_vf(r, X).T
3  display(Math(f" \\text{{Jacobian matrix of reversed r = }}{latex(J)}"))
```

Jacobian matrix of reversed $\mathbf{r} = \begin{bmatrix} -1 & 0 \\ 0 & -1 \end{bmatrix}$

In this case, the two eigenvalues are negative at any point in the entire field, indicating that this vector field is attractive. If the vector field represents fluid flow, the fluid at any point flows into that point. If the vector field represents a force field, a particle at any point is drawn toward that point.

For a negated position (or inward radial) field, the vector field radiates toward its center, as shown in Fig. 3.3.

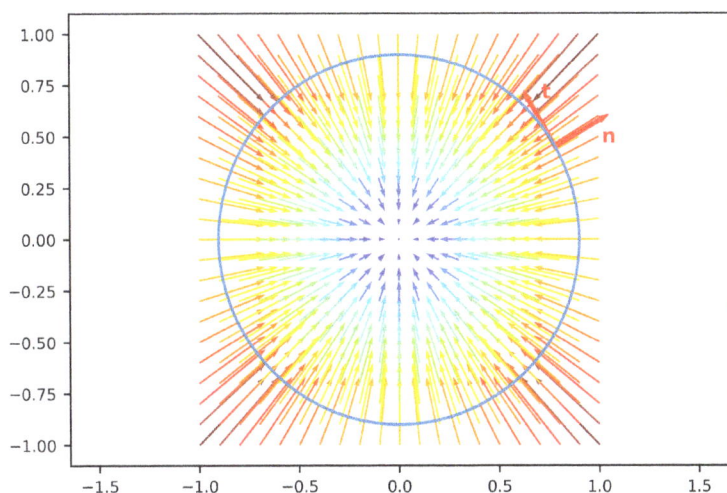

Figure 3.3. The negated position vector field, which radiates inward.

The negated position field is plotted using the following snippet:

```
1  fx = lambda x, y: -x + y*0      # inward radial field, functons of x & y
2  fy = lambda x, y: x*0 - y
3  plt = plotField(a, b, fx, fy, 'positionFieldn')
4  #plt.show()
```

3.1.3 *Spin vector field*

The spin vector **s** is a rotation vector that orbits around the origin $(0,0)$. Its direction follows the counterclockwise convention, and the spin vector with respect to the (z)-axis is expressed as

$$\mathbf{s} = -y\mathbf{i} + x\mathbf{j} = \begin{bmatrix} -y \\ x \end{bmatrix} \tag{3.5}$$

In this case, $s_x = -y$ and $s_y = x$. The direction of **s** is in the circumferential direction and hence is perpendicular (or orthogonal) to the position vector **r**. Therefore, the dot product of **r** and **s** vanishes:

$$\mathbf{r} \cdot \mathbf{s} = x(-y) + yx = 0 \tag{3.6}$$

The spin vector fields in 3D can be expressed as

$$\mathbf{s}_x = \begin{bmatrix} 0 \\ -z \\ y \end{bmatrix}; \quad \mathbf{s}_y = \begin{bmatrix} z \\ 0 \\ -x \end{bmatrix}; \quad \mathbf{s}_z = \begin{bmatrix} -y \\ x \\ 0 \end{bmatrix} \tag{3.7}$$

In this case, we have three possible spin fields, $\mathbf{s}_x$, $\mathbf{s}_y$, and $\mathbf{s}_z$, each spinning around the x-, y-, and z-axes, respectively. It is straightforward to confirm that each of these three is orthogonal to $\mathbf{r}$. The following snippet illustrates this confirmation:

```
1  x, y, z = symbols('x, y, z', real=True)        # define variables
2  r = sp.Matrix([x, y, z])                    # the 3D position vector
3  sx = sp.Matrix([0, -z, y])                     # the 3D spin vector
4
5  display(Math(f" \\text{{Is r and s$_x$ orthogonal? }}\
6                          {latex(r.dot(sx)==0)}"))
```

Is r and $\mathbf{s}_x$ orthogonal? True

Let us plot the 2D spin field:

```
1  fx = lambda x, y: x*0 - y                 # define the spin field
2  fy = lambda x, y: x + y*0
3
4  plt = plotField(a, b, fx, fy, 'spinField')        # plot name as str
5  #plt.show()
```

The most significant feature of a spin vector field is that it circulates around a point, as shown in Fig. 3.4. Hence, it has circulation but no radiation and is divergence-free. It is also called a solenoidal vector field. If it represents fluid flow, it is called incompressible flow because its volume change (measured by the divergence) is zero:

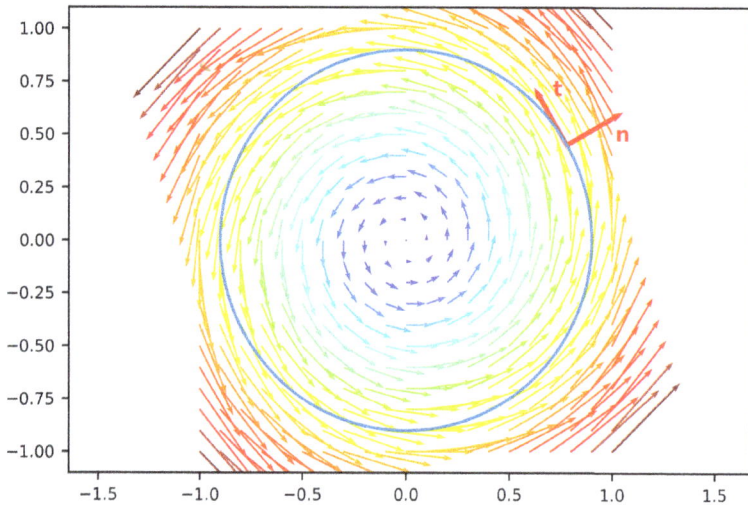

Figure 3.4.   The spin vector field, which circulates around a point.

```
1  X = sp.Matrix([x, y])                        # the 3D cooridnate vector
2  sx2D = sp.Matrix([-y, x])                     # the 3D position vector
3
4  display(Math(f" \\text{{curl of the spin vector = }}\
5                       {latex(gr.curl_vf(sx2D, X).simplify())}"))
6  display(Math(f" \\text{{divergence of the spin vector = }}\
7                       {latex(gr.div_vf(sx2D, X))}"))
```

Curl of the spin vector $= 2$

Divergence of the spin vector $= 0$

This time, we find that the curl of the spin vector is a constant 2. This is expected because the entire field is rotating (circulating) in the tangent direction with respect to $(0,0)$. The rate of rotation is 2.

On the other hand, the divergence of the spin vector field is zero. This implies that the spin field is not diverging (spreading in or out). This also means that if this field exists, nothing would be leaving any circular curve centered at $(0,0)$. This is another example of a mathematical description of a conservation law.

Let us compute the Jacobian matrix and its eigenvalues for this spin field:

```
1  J = gr.grad_vf(sx2D, X).T
2  display(Math(f" \\text{{Jacobian matrix of s$_{{{x2D}}}$ = }}{latex(J)}"))
3  eigsJ = J.eigenvals()
4  display(Math(f" \\text{{Eigenvalues of J = }}{latex(eigsJ)}"))
```

Jacobian matrix of $\mathbf{s}_{x2D} = \begin{bmatrix} 0 & -1 \\ 1 & 0 \end{bmatrix}$

Eigenvalues of $\mathbf{J} = \{-i : 1, \ i : 1\}$

It is found that the eigenvalues of the Jacobian matrix $\mathbf{J}$ are complex-valued. This means that every point in the field is a spiral point, a typical feature of spin fields. Due to these unique characteristics of the position and spin vector fields, they will be frequently used as basis vector fields for various purposes in our study.

A spin field can have an arbitrarily varying amplitude in the radial direction, indicating that the rate of circulation is not necessarily constant. For example, the following (normalized) spin field has a constant amplitude:

$$\mathbf{s}_n = \begin{bmatrix} \dfrac{-y}{\sqrt{x^2+y^2}} \\ \dfrac{x}{\sqrt{x^2+y^2}} \end{bmatrix} \tag{3.8}$$

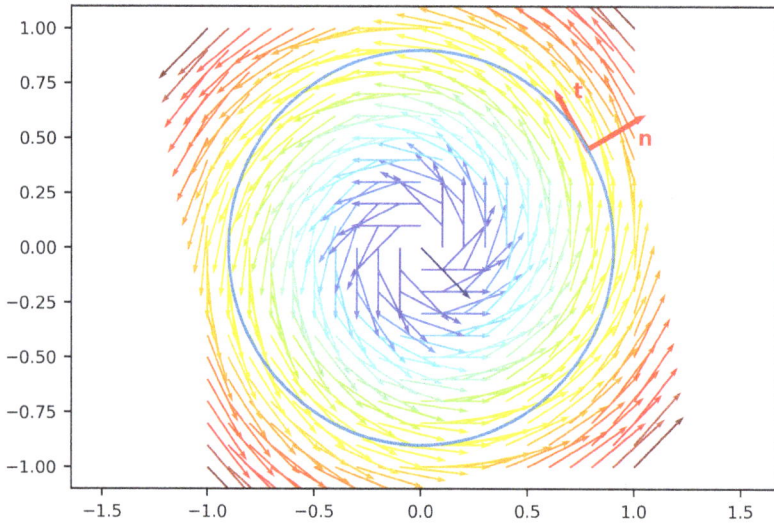

Figure 3.5.   The spin vector field with constant amplitude, which circulates around a point.

The following snippet plots show the normalized spin field:

```
1  snx = lambda x, y: (x*0 - y)/np.sqrt(x**2+y**2) # define the spin field
2  sny = lambda x, y: (x + y*0)/np.sqrt(x**2+y**2)
3  plt = plotField(a, b, snx, sny, 'spin_rField')        # plot name as str
4  #plt.show()
```

Figure 3.5 shows spin field with constant amplitude. The curl and divergence of this spin field is computed using the following:

```
1  X = sp.Matrix([x, y])                           # the 2D cooridnate vector
2  s_n = sp.Matrix([-y/sqrt(x**2+y**2),x/sqrt(x**2+y**2)]) # a spin vector
3
4  display(Math(f" \\text{{curl of the normalized spin vector = }}\
5                      {latex(gr.curl_vf(s_n, X).simplify())}"))
6  display(Math(f" \\text{{divergence of the normalized spin vector = }}\
7                      {latex(gr.div_vf(s_n, X))}"))
```

Curl of the normalized spin vector $= \dfrac{1}{\sqrt{x^2 + y^2}}$

Divergence of the normalized spin vector $= 0$

In this case, the curl of the spin vector is $\frac{1}{r}$. The rate of circulation approaches infinity as r approaches zero, and it approaches zero as r approaches infinity. The divergence remains zero.

Let us compute the eigenvalues of its Jacobian matrix:

```
1  J = sp.simplify(gr.grad_vf(s_n, X).T)     # Generate the Jacobian matrix
2  display(Math(f" \\text{{Jacobian matrix of s$_{{x2D}}$ = }}{latex(J)}")
3  eigsJ = J.eigenvals()
4  display(Math(f" \\text{{Eigenvalues of J = }}{latex(eigsJ)}"))
```

Jacobian matrix of $\mathbf{s}_{x2D} = \begin{bmatrix} \dfrac{xy}{(x^2+y^2)^{\frac{3}{2}}} & -\dfrac{x^2}{(x^2+y^2)^{\frac{3}{2}}} \\ \dfrac{y^2}{(x^2+y^2)^{\frac{3}{2}}} & -\dfrac{xy}{(x^2+y^2)^{\frac{3}{2}}} \end{bmatrix}$

Eigenvalues of $\mathbf{J} = \{0:2\}$

This time, we obtain two zero eigenvalues for the entire field. This means that the field is neither purely repulsive, attractive, saddle nor spiral. Instead, this field exhibits characteristics of both radial and spin fields. It demonstrates spinning behavior and also has a potential component.

3.1.4 *An arbitrary vector field*

In general, a vector field encountered in science and engineering can be arbitrary. It is typically a combination of radial and spin components. Following is an arbitrarily constructed vector field:

$$\mathbf{f} = \begin{bmatrix} -xy \\ x+y \end{bmatrix} \tag{3.9}$$

The field vectors are plotted using the same code:

```
1  from matplotlib import cm
2  plt.rcParams.update({'font.size': 8})
3  plt.rcParams['figure.dpi'] = 500
4  a, b = 1., 1.                            # domain [-a, a]×[-b, b] for plots
5
6  fx = lambda x,y: -x*y    #x*y-y              # one may try alternatives
7  fy = lambda x,y:  x+y    #x**2 + y**2 + x
8
9  plt = plotField(a, b, fx, fy, 'generalField')        # plot name as str
10 #plt.show()
```

This vector field radiates in various directions and also includes circulations at different locations, as shown in Fig. 3.6. This plot is only for 2D; one can imagine the complexity of a general vector field in 3D space.

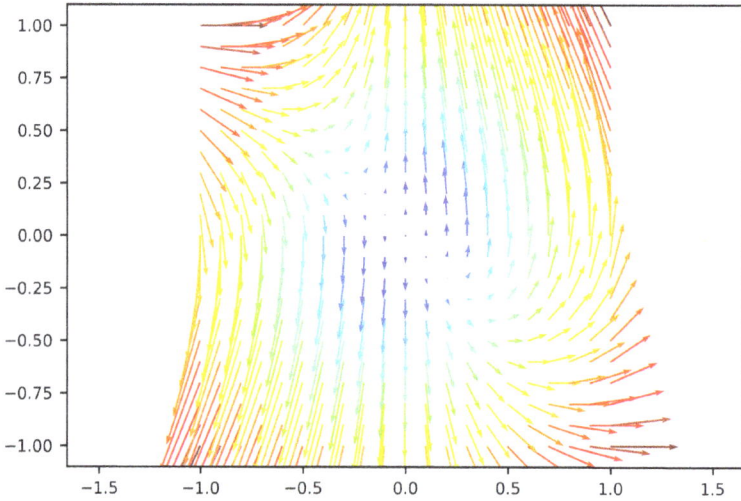

Figure 3.6. An arbitrary vector field, which radiates in different directions and is mixed with circulations.

For these types of fields, there are generally both divergence and curl. Let us confirm with the following:

```
1  x, y = symbols('x, y ', real=True)                    # define variables
2  X = sp.Matrix([ x, y])                       # the 2D cooridnate vector
3  f = sp.Matrix([-x*y, x+y])                      # a 2D arbitrary vector
4  display(Math(f" \\text{{curl of the arbitrary field = }}\
5                          {latex(gr.curl_vf(f, X).simplify())}"))
6  display(Math(f" \\text{{divergence of the arbitrary field = }}\
7                          {latex(gr.div_vf(f, X))}"))
```

Curl of the arbitrary field $= x + 1$

Divergence of the arbitrary field $= 1 - y$

For this arbitrary vector field, we found both curl and divergence; these are functions of the coordinates, implying that both are changing with respect to their location. This also means that if this field exists, there must be distributed sources and it flows both in and out, with varying directions.

In fact, any vector field that is sufficiently smooth in simply connected three-dimensional space can be resolved into the sum of a curl-free and a divergence-free vector field. This is known as the **Helmholtz decomposition**, which will be discussed in great detail in the final section of this chapter.

Let us compute the Jacobian matrix and its eigenvalues for the field given in Eq. (3.9):

```
1  J = sp.simplify(gr.grad_vf(f, X).T)       # Generate the Jacobian matrix
2  display(Math(f" \\text{{Jacobian matrix J = }}{latex(J)}"))
3  eigsJ = J.eigenvals()
4  display(Math(f" \\text{{Eigs =}}{latex(eigsJ)}"))
```

Jacobian matrix $\mathbf{J} = \begin{bmatrix} -y & -x \\ 1 & 1 \end{bmatrix}$

$$\text{Eigs} = \left\{ -\frac{y}{2} - \frac{\sqrt{-4x + y^2 + 2y + 1}}{2} + \frac{1}{2} : 1, \right.$$

$$\left. -\frac{y}{2} + \frac{\sqrt{-4x + y^2 + 2y + 1}}{2} + \frac{1}{2} : 1 \right\}$$

For this arbitrary vector field, the eigenvalues can be real or complex, depending on the location (x, y). This means that a point in the field may exhibit repulsive, attractive, saddle, or spiral behavior.

Therefore, different types of vector fields can exhibit various behaviors, and their effects need to be properly measured. This is often achieved using a number of **mathematical operators** that involve partial derivatives and various types of integrations for the vector fields. Several important **theorems** have also been developed to capture the features of vector fields. The following chapters will discuss these in detail.

3.2 Work done by a vector field along a curve

As shown in these plots of vector fields, a vector function has both direction and amplitude, and both vary within the domain. The effects of the vector field naturally depend on the location and direction of the vectors, whether they act along a curve, across a surface, or within a domain.

Here, we consider two important cases: the effect of a vector field along a curve and the effect across a curve defined in the domain. Both cases are significant in applications in science and engineering.

3.2.1 *Effects of a vector field along a curve*

Consider an arbitrary vector field $\mathbf{V}$ in a 2D domain $\mathcal{D}$ defined in Cartesian coordinates (x, y):

$$\mathbf{V} = \begin{bmatrix} V_x(x, y) \\ V_y(x, y) \end{bmatrix} = \begin{bmatrix} M(x, y) \\ N(x, y) \end{bmatrix} \tag{3.10}$$

It has two component functions, $V_x(x, y)$ and $V_y(x, y)$, which are often denoted in the literature as $M(x, y)$ and $N(x, y)$, respectively. The effect (E) of the vector field along a given curve $\mathcal{C}$ in the domain $\mathcal{D}$ in the tangential direction can be measured using a line integral:

$$E = \int_{\mathcal{C}} \mathbf{V} \cdot \mathbf{t}\, ds \tag{3.11}$$

where $\mathbf{t}$ is the **unit tangent vector** at point (x, y) along $\mathcal{C}$, as shown in Fig. 3.7. Here, we require the integrand to be integrable. Both $\mathbf{V}$ and $\mathbf{t}$ are vectors in 2D space: The former is the given vector field and the latter depends on the curve chosen. Their dot product produces a scalar function that measures the effect of $\mathbf{V}$ along the direction of $\mathbf{t}$ at a point on $\mathcal{C}$. The integration sums up all these effects along the curve $\mathcal{C}$ and produces a net effect. The position vector $\mathbf{r}$ is used here to conveniently locate the points on $\mathcal{C}$.

Equation (3.11) reveals the following:

1. If $\mathbf{V}$ at (x, y) is perpendicular to $\mathbf{t}$, the effect E of the vector field along the curve is zero.
2. If $\mathbf{V}$ is in the same direction as $\mathbf{t}$, the effect will be simply $\|\mathbf{V}\|ds$, and the value will be positive.

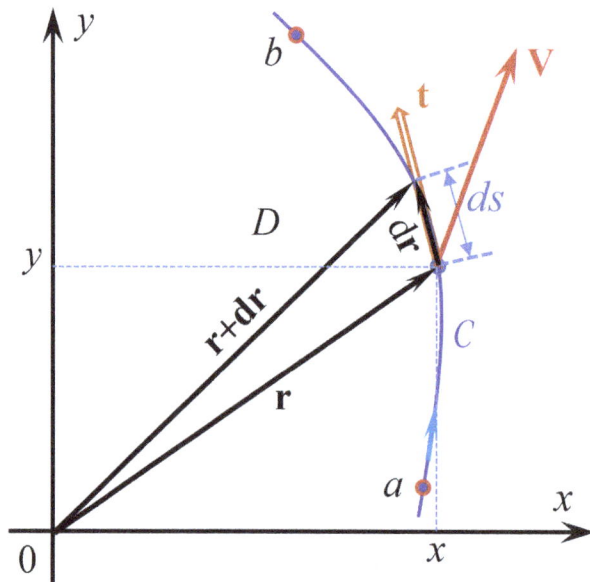

Figure 3.7. A field vector $\mathbf{V}$ on a curve $\mathcal{C}$ in a 2D plane and the position vector $\mathbf{r}$ at point (x, y).

3. If $\mathbf{V}$ is in the direction opposite to $\mathbf{t}$, the effect will be $-\|\mathbf{V}\| ds$, and the value will be negative.
4. In any case, only the component of $\mathbf{V}$ in the direction of $\mathbf{t}$ will have an effect. The dot product given in Eq. (3.11) captures this automatically at different locations on the curve.

The effect E will be a scalar value resulting from the integration of the dot product along $\mathcal{C}$.

As shown in Fig. 3.7, due to the location change of ds at point (x, y) along $\mathcal{C}$, the position vector $\mathbf{r}$ should have a corresponding change $d\mathbf{r}$. We thus have $\mathbf{t} ds = d\mathbf{r}$, and Eq. (3.11) becomes

$$E = \int_{\mathcal{C}} \mathbf{V} \cdot d\mathbf{r} \tag{3.12}$$

The position vector $\mathbf{r}$ is given in Eq. (3.2), and hence $d\mathbf{r}$ becomes

$$\mathbf{t} ds = d\mathbf{r} = \begin{bmatrix} dx \\ dy \end{bmatrix} \tag{3.13}$$

Using Eqs. (3.10) and (3.13), Eq. (3.12) can be rewritten as

$$E = \int_{\mathcal{C}} V_x \, dx + V_y \, dy \tag{3.14}$$

This allows us to evaluate the effects using the components of the given vector field.

In general, $\mathbf{r}$ can be parameterized (see Section 4.14 in Ref. [4]) as $\mathbf{r}(t) = [x(t) \; y(t)]^{\top}$, and hence Eq. (3.13) can be written as

$$\mathbf{t} ds = d\mathbf{r} = \begin{bmatrix} \frac{\partial x}{\partial t} \\ \frac{\partial y}{\partial t} \end{bmatrix} dt \tag{3.15}$$

These identities are useful in carrying out integrals along a curve $\mathcal{C}$, which will be demonstrated in the example sections.

3.2.2 *Work done by a force field along a curve*

Now, assume the vector field $\mathbf{V}$ is a force field $\mathbf{F}$ defined in Cartesian coordinates (x, y), and hence it has two component functions F_x and F_y:

$$\mathbf{F} = \begin{bmatrix} F_x \\ F_y \end{bmatrix} \tag{3.16}$$

The effect of the force vector field will be the work done W, which is given by Eq. (3.14) by replacing V_x and V_y with F_x and F_y, respectively:

$$W = \int_C \mathbf{F} \cdot \mathbf{t} \, ds = \int_C \mathbf{F} \cdot d\mathbf{r} = \int_C F_x \, dx + F_y \, dy \qquad (3.17)$$

It is clear that the work done is contributed by these two force components over the differential distance in their respective directions, which may be a familiar fact to readers.

Let us look at an example.

3.2.3 *Example: Work by a gravity field*

Consider a gravity field near the Earth's surface. The force vector acting on a small ball with constant mass m is given by

$$\mathbf{F} = \begin{bmatrix} 0 \\ -mg \end{bmatrix} \qquad (3.18)$$

where g is the acceleration due to gravity. Since we are considering the field near the Earth's surface, it can be treated as a 2D problem. The direction of the force is straight downward, and the gravity value can be assumed to be a constant, 9.8 m/s^2.

Assume the ball falls from a height of $H = 100 \text{ m}$. Let us compute the work done by the gravity field. We consider two paths: a straight downward path C_1 and a semi-circular path C_2 with radius $R = H/2$ (caused by some horizontal forces), as shown in Fig. 3.8.

Solution:

Case I: Along the straight downward path C_1. In this case, by introducing the parameter t: $x = 0 \times t$, $y = t$, the position vector is

$$\mathbf{r} = \begin{bmatrix} 0 \\ y \end{bmatrix} = \begin{bmatrix} 0 \\ t \end{bmatrix} \qquad (3.19)$$

and d$\mathbf{r}$ becomes

$$d\mathbf{r} = \begin{bmatrix} 0 \\ 1 \end{bmatrix} dt \qquad (3.20)$$

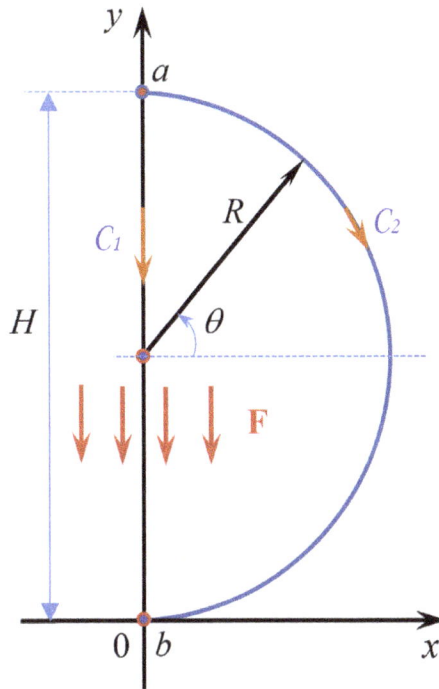

Figure 3.8. Work done by the vertical gravity force field along two curves in a 2D plane: vertical down C_1 and semi-circular curve C_2.

With these defined, the work done can be found using Eq. (3.17). The code is as follows:

```
1  x, y, t = symbols('x, y, t', real=True)                    # define variables
2  m, g, H = symbols('m, g, H', positive=True)
3  Fx, Fy = symbols('F_x, F_y', real=True)
4  H = 100  # (m)
5  Gx = 0*g                                        # gravity force in the x-direction
6  Gy =-m*g                        # gravity force in the y-direction, (kg m/s^2 = N)
7
8  Gv = sp.Matrix([Gx, Gy])                                    # force vector
9  x_ = 0*t                                              # x_: x given in t
10 y_ = t                                                # y_: y given in t
11 dic = {x:x_, y:y_}                                    # for later substitution
12 dRdt = sp.Matrix([x_.diff(t), y_.diff(t)])    # x = 0*t; y = t in [H, 0]
13
14 # Use the components formula:
15 W_C1 = integrate(Gy.subs(dic),(t, H, 0))      # only need the y-component
16 display(Math(f" \\text{{Work done (Nm) by the gravity force along C1, \
17                       via components formula = }} {latex(W_C1)} "))
18
19 # Use the dot-product formula:
20 W_C1 = integrate((Gv.T@dRdt).subs(dic), (t, H, 0))[0]
21 display(Math(f" \\text{{Work done (Nm) by the gravity force along C1, \
22                       via dot-product formula = }} {latex(W_C1)} "))
```

Work done (Nm) by the gravity force along C_1 via components formula =
100 gm

Work done (Nm) by the gravity force along C_1 via dot-product formula =
100 gm

Case II: Work done along the semi-circular path C_2. In this case, we
introduce the parameter θ, and the position vector is given by (see Fig. 3.8)

$$\mathbf{r} = \begin{bmatrix} R\cos\theta \\ y_0 + R\sin\theta \end{bmatrix} \tag{3.21}$$

where $R = \frac{H}{2}$ and $y_0 = \frac{H}{2}$ (which is not at the origin). Thus, $\mathbf{dr}$ becomes

$$\mathbf{dr} = \begin{bmatrix} -R\sin\theta \\ R\cos\theta \end{bmatrix} d\theta \tag{3.22}$$

The work done can be calculated using the dot-product formula in
Eq. (3.17). The code is as follows:

```
1  R, y0 = symbols('R, y0', positive=True)
2  θ = symbols('θ ', real=True)
3  x_ = R*sp.cos(θ)                          # Cartesian-polar mapping
4  y_ = y0 + R*sp.sin(θ)
5
6  dRdt = sp.Matrix([x_.diff(θ), y_.diff(θ)])        # θ in [pi/2,-pi/2]
7  dic = {x:x_, y:y_, R:50}
8  W_C2 = sp.integrate((Gv.T@dRdt).subs(dic), (θ, sp.pi/2, -sp.pi/2))[0]
9  display(Math(f" \\text{{Work done (Nm) by the gravity force along C2,\
10                     via dot-product formula = }} {latex(W_C2)}"))
```

Work done (Nm) by the gravity force along C_2 via dot-product formula =
100 gm

We have obtained the exact same result. The work done by the gravitational force field is path-independent. If we integrate over the interval
$[\pi/2, 3\pi/2]$ for the parameter θ, which corresponds to the semi-circle on the
left, we will also get the same value:

```
1  W_C3 = sp.integrate((Gv.T@dRdt).subs(dic), (θ, sp.pi/2, 3*sp.pi/2))[0]
2  display(Math(f" \\text{{Work done (Nm) by the gravity force along C3,\
3                     (θ from π/2 to 3π/2) = }} {latex(W_C3)}"))
```

Work done (Nm) by the gravity force along C_3 (θ from $\pi/2$ to $3\pi/2$) =
100 gm

Furthermore, if we connect C_1 and C_2 to form a closed path in the counterclockwise direction, the C_2 segment will contribute a negative value. The sum of the results will be zero. This indicates that if a line integral is path-independent, then the integral over a closed curve will also be zero.

It is interesting to check the eigenvalues of this 2D gravitational field:

```
1  J = sp.simplify(gr.grad_vf(Gv, X).T)    # Generate the Jacobian matrix
2  display(Math(f" \\text{{Jacobian matrix of the gravity field = }}\
3                                          {latex(J)}"))
4  eigsJ = J.eigenvals()
5  display(Math(f" \\text{{Eigenvalues of J = }}{latex(eigsJ)}"))
```

Jacobian matrix of the gravity field $= \begin{bmatrix} 0 & 0 \\ 0 & 0 \end{bmatrix}$

Eigenvalues of $\mathbf{J} = \{0 : 2\}$

It is observed that the Jacobian matrix is a zero matrix. This is because this 2D idealized gravitational field is constant, and hence its component functions do not change with the coordinates. Consequently, the eigenvalues must also be zero.

The above analysis captures important features of the gravitational field. Not all vector fields possess this property. Let us now examine the spin field.

3.2.4 *Example: Work done by a spin field along a curve*

Consider a spin field defined by Eq. (3.5) as the force field. We have

$$\mathbf{F} = \begin{bmatrix} -y \\ x \end{bmatrix} \tag{3.23}$$

We will compute the work done by this special force field, considering the same two paths as in the previous example.

Solution: We will use Eq. (3.17) and the same code to compute the work done. First, consider Case I:

```
 1  Fx = -y                                    # force in the x-direction
 2  Fy =  x                    # force in the y-direction, (kg m/s^2 = N)
 3  Fv = sp.Matrix([Fx, Fy])                            # force vector
 4
 5  x_ = 0*t                                          # use parameter t
 6  y_ = t
 7  dic = {x:x_, y:y_, R:50}
 8  dRdt = sp.Matrix([x_.diff(t), y_.diff(t)])
 9
10  # use the two components:
11  W_C1 = integrate(Fx.subs(dic),(x,0,0))+integrate(Fy.subs(dic),(y,H,0))
12  display(Math(f" \\text{{Work done (Nm) by the spin field along C1,\
13                      via component formula = }} {latex(W_C1)}"))
14  # use the dot-product:
15  W_C1 = integrate((Fv.T@dRdt).subs(dic), (y, H, 0))[0]
16  display(Math(f" \\text{{Work done (Nm) by the spin field along C1,\
17                      via dot-product = }} {latex(W_C1)}"))
```

Work done (Nm) by the spin field along C_1 via component formula $= 0$

Work done (Nm) by the spin field along C_1 via dot-product $= 0$

This time, we obtained a result of zero for the work done. This makes sense because the spinning force field along the vertical line acts in the horizontal direction. Such spinning forces cannot perform work in this context.

Let us now find the result for Case II:

```
 1  R, y0 = symbols('R, y0', positive=True)
 2  θ = symbols('θ ', real=True)
 3  x_ = R*sp.cos(θ)           # use parameter θ; Cartesian-polar mapping
 4  y_ = y0 + R*sp.sin(θ)
 5
 6  dRdt = sp.Matrix([x_.diff(θ), y_.diff(θ)])          # θ in [pi/2,-pi/2]
 7  dic = {x:x_, y:y_, R:50}
 8  W_C2 = sp.integrate((Fv.T@dRdt).subs(dic), (θ, sp.pi/2, -sp.pi/2))[0]
 9  display(Math(f" \\text{{Work done (Nm) by the spin field along C1,\
10                      via dot-product = }} {latex(W_C2)}"))
```

Work done (Nm) by the spin field along C_1 via dot-product $= -50\pi R$

In this case, the spin force field has done some work. The work is negative because the directions of the spin force are largely opposite to the tangent direction of the semi-circular path (which is downward). If we integrate over the interval $[\pi/2, 3\pi/2]$, which corresponds to the semi-circle on the left, we will obtain a positive value:

```
 1  W_C3 = sp.integrate((Fv.T@dRdt).subs(dic), (θ, sp.pi/2, 3*sp.pi/2))[0]
 2  display(Math(f" \\text{{Work done (Nm) by the spin field along C3,\
 3                      (θ from π/2 to 3π/2) = }} {latex(W_C3)}"))
```

Work done (Nm) by the spin field along C_3 (θ from $\pi/2$ to $3\pi/2$) $= 50\pi R$

3.3 Work done by a vector field along a 3D curve

3.3.1 *Formulation*

We discussed curve integrations in 3D in Ref. [4]. An effective technique for such integrals is to parameterize the curve by introducing a parameter. This approach closely resembles the 2D cases discussed above, with the formulas serving as straightforward extensions from the 2D case.

Equation (3.11) still holds for 3D, but d**r** becomes

$$\mathbf{t}ds = d\mathbf{r} = \begin{bmatrix} dx \\ dy \\ dz \end{bmatrix} \tag{3.24}$$

In general, **r** can be parameterized (see Section 4.14 in Ref. [4]) as $\mathbf{r}(t) = [x(t) \ y(t) \ z(t)]^{\top}$, and hence Eq. (3.13) can be written as

$$\mathbf{t}ds = d\mathbf{r} = \begin{bmatrix} \frac{\partial x}{\partial t} \\ \frac{\partial y}{\partial t} \\ \frac{\partial z}{\partial t} \end{bmatrix} dt \tag{3.25}$$

Equation (3.12) can be rewritten as

$$E = \int_C V_x dx + V_y dy + V_z dz = \int_C \mathbf{V} \cdot d\mathbf{r} \tag{3.26}$$

3.3.2 *Example: Work done by a gravity field along a helix*

Consider an elliptic helix in 3D. This example computes the work done by the gravity field studied in the previous example, but along an elliptic helix. The helix can be parameterized using a parameter θ:

$$x = a\cos(\theta) - a$$
$$y = b\sin(\theta) \tag{3.27}$$
$$z = h\theta + H$$

where a is the semi-major axis, b is the semi-minor axis of the ellipse, h is the pitch (the z-increment per turn), and H is the total height of the helix. To provide a better visualization of such a curve, the following snippet plots a typical helix:

```
1   # Parameters for the elliptic helix
2   a = 8;   b = 4                           # Semi-major and Semi-minor axes
3   h = 1                                    # Pitch (z-increment/turn)
4   H = 100                                  # total height
5
6   # Parametric equations for an elliptic helix
7   θ = np.linspace(0, 4*np.pi, 1000)        # Parameter θ for the curve
8   θU = 4*np.pi
9   h = -H/θU
10  x = a*np.cos(θ)-a;   y = b * np.sin(θ)
11  z = h*θ + H
12
13  plt.ioff()# plt.ion()
14  fig = plt.figure()#figsize=(7, 4))                    # in inches
15  ax = fig.add_subplot(111, projection='3d')
16  ax.plot(x, y, z, color='b', lw=2)
17  ax.scatter(0,0,H, color='r')                          # starting point
18  ax.scatter(0,0,0, color='g')                          # ending point
19
20  ax.set_xlabel('X axis', labelpad=-9); ax.tick_params(axis='x', pad=-5)
21  ax.set_ylabel('Y axis', labelpad=-9); ax.tick_params(axis='y', pad=-5)
22  ax.set_zlabel('Z axis', labelpad=-5); ax.tick_params(axis='z', pad=-2)
23  plt.savefig('imagesVC/ellipticHelix.png', dpi=500)    # save to file
24  #plt.show()
```

A typical elliptic helix is plotted in Fig. 3.9.

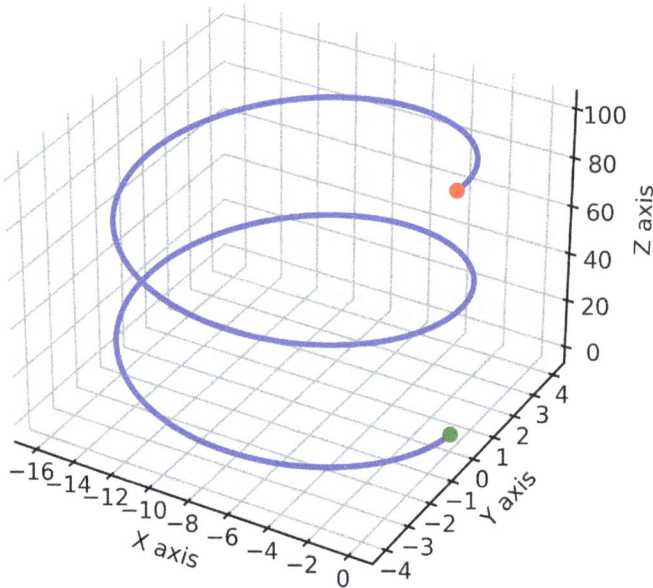

Figure 3.9.   An elliptic helix in 3D space where there may be a vector field.

Solution: The work done by the gravity field from the red point at a height of 100 m to the ground at 0 m can be computed using the following snippet. This approach is an extension of the code used for the 2D cases:

```
1  x, y, z, θ = symbols('x, y, z, θ', real=True)          # define variables
2  m, g, H = symbols('m, g, H', positive=True)
3  Fx, Fy, Fz = symbols('F_x, F_y, F_z', real=True)
4  H = 100  # (m)
5  Gx = 0*g                                 # gravity force in the x-direction
6  Gy = 0*g                                        # the y-direction
7  Gz =-m*g                                 # gravity force in the y-direction
8
9  Gv = sp.Matrix([Gx, Gy, Gz])                             # 3D force vector
10 # Parametric equations for an elliptic helix
11 θU = 4*np.pi
12 h = -H/θU
13 x_ = a * cos(θ) - a
14 y_ = b * sin(θ)
15 z_ = h * θ + H
16 dic = {x:x_, y:y_, z:z_}                        # for later substitution
17 dRdθ = sp.Matrix([x_.diff(θ), y_.diff(θ), z_.diff(θ)])
18
19 # Use the dot-product formula:
20 W_Helix = sp.integrate((Gv.T@dRdθ).subs(dic), (θ, 0, θU))[0]
21 display(Math(f" \\text{{Work done (Nm) by the gravity field along helix
22                                 = }} {latex(W_Helix)}"))
```

Work done (Nm) by the gravity field along helix $= 100.0\,\mathrm{gm}$

We found that the work done is the same along both the straight downward path C_1 and the semi-circular path C_2.

3.3.3 *Example: Work done by spin fields along a helix*

Now, consider the same elliptic helix in 3D. This example computes the work done by the spin field defined in Eq. (3.7).

Solution: We can use the same snippet to compute the results:

```
1  sx = sp.Matrix([0, -z, y])           # 3D spin field with respect to x-axis
2  #sy = sp.Matrix([z, 0, -x])                          # w.r.t y-axis
3  #sz = sp.Matrix([-y, x, 0])                          # w.r.t z-axis
4
5  dRdθ = sp.Matrix([x_.diff(θ), y_.diff(θ), z_.diff(θ)])
6  integrand = (sx.T@dRdθ).subs(dic)
7  display(Math(f" \\text{{The integrand along the helix = }} \
8                                 {latex(integrand.evalf(4))}"))
9  W_Helix = sp.integrate((sx.T@dRdθ).subs(dic), (θ, 0, θU))[0]
10 display(Math(f" \\text{{Work done by the spin field along helix = }} \
11                                 {latex(W_Helix)}"))
```

The integrand along the helix $=$ $[-4.0 \cdot (100.0 - 7.958\theta)\cos(\theta) - 31.83\sin(\theta)]$

Work done by the spin field along helix $= 0$

As observed, the work done is zero despite the integrand being nonzero. This result also holds true for the other two spin fields. Readers can uncomment the lines defining these other spin fields and run the code to verify this. The reason for this outcome is that the helix effectively forms a closed path when projected onto the x–y plane. This result will always be true if the ending θ is an even multiple of π. However, if we change the ending point to an odd multiple, such as 3π, we will obtain a nonzero value. Let us confirm this with the following:

```
1 W_Helix = sp.integrate((sx.T@dRdθ).subs(dic), (θ, 0, 3*sp.pi))[0]
2 display(Math(f" \\text{{Work done by the spin field along helix = }} \
3                                        {latex(W_Helix)}"))
```

Work done by the spin field along helix $= -127.323954473516$

We obtained a nonzero value, as anticipated.

From the examples above, we observe that the work done by the gravity field is path-independent, while the work done by the spin field is path-dependent. This raises an important question: What is the condition for a field to exhibit path-independent work? The answer lies in a special type of field known as a **gradient field**.

3.4 Gradient fields

3.4.1 *Definition and potential functions*

As discussed in Section 2.5, the gradient of a scalar continuously differentiable function $\phi(x, y, z)$ is a vector field with three component functions, denoted as $\nabla\phi$. This field is known as the **gradient field** and is expressed as

$$\nabla\phi = \frac{\partial\phi}{\partial x}\mathbf{i} + \frac{\partial\phi}{\partial y}\mathbf{j} + \frac{\partial\phi}{\partial z}\mathbf{k} = \begin{bmatrix} \frac{\partial\phi}{\partial x} \\ \frac{\partial\phi}{\partial y} \\ \frac{\partial\phi}{\partial z} \end{bmatrix} \tag{3.28}$$

3.4.2 *Potential function contours*

It is important to note the following:

> The vector field $\nabla\phi$ is perpendicular to the contour lines $\phi(x, y, z) =$ constant at any point in the domain.

This can be confirmed through the following procedure, considering a 2D case.

Figure 3.10 schematically shows three contour lines of a 2D function ϕ, with $\phi(x, y) = c_1$, $\phi(x, y) = c_2$, and $\phi(x, y) = c_3$, where c_i (for $i = 1, 2, \ldots$) are constants.

At a point (x, y) on the contour line $\phi(x, y) = c_1$, the tangent direction **t** aligns with that of **dr**, which is given by $[dx; dy]^\top$. The dot product of $\nabla\phi$ and **dr** is expressed as follows:

$$\nabla\phi \cdot d\mathbf{r} = \frac{\partial\phi}{\partial x}dx + \frac{\partial\phi}{\partial y}dy = d\phi = 0 \tag{3.29}$$

In the last step, we utilized the fact that ϕ remains constant along the contour line, making its total differential with respect to **dr** equal to zero. Equation (3.29) indicates that $\nabla\phi$ and **dr** are orthogonal at any point (x, y) on a contour line.

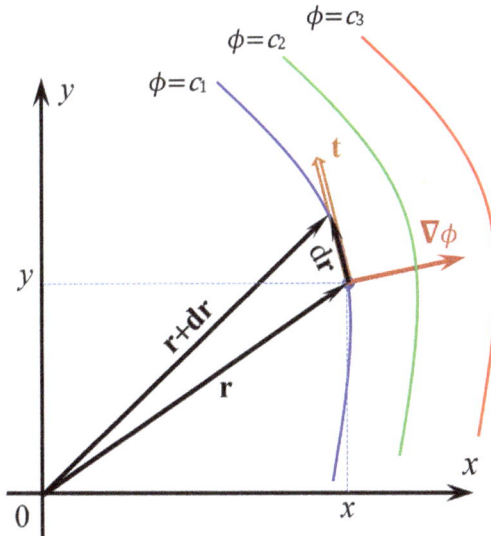

Figure 3.10. Sample contour lines of a scalar potential function ϕ.

The length $\|\nabla\phi\|$ represents the rate of change of $\phi(x, y, z)$ in the direction of $\nabla\phi$. This is the direction in which ϕ changes most rapidly. This concept forms the basis of gradient ascent (or descent) algorithms, where movement occurs in the direction of the gradient to maximize (or minimize) the function [4, 5].

A gradient field is a special type of vector field characterized by its component functions being the partial derivatives of a scalar function ϕ. Specifically, the components of the gradient field are as follows:

$$V_x = \frac{\partial\phi}{\partial x}; \quad V_y = \frac{\partial\phi}{\partial y}; \quad V_z = \frac{\partial\phi}{\partial z} \tag{3.30}$$

This implies that the vector field $\mathbf{V}$ can be expressed as the gradient of a scalar **potential function** ϕ. In other words, if such a potential function ϕ exists, then $\mathbf{V}$ is indeed a gradient field.

In the 2D case, the vector field components $\frac{\partial\phi}{\partial x}$ and $\frac{\partial\phi}{\partial y}$ represent the gradient of a scalar potential function ϕ, describing the 2D gradient field.

Clearly, not all vector fields are gradient fields; this depends on the existence of the potential function.

3.4.3 *Work by a gradient field: A conservative field*

Let us compute the work done by a gradient field in 2D along a curve $\mathcal{C}$ using Eq. (3.17):

$$W = \int_{\mathcal{C}} \mathbf{V} \cdot d\mathbf{r} = \int_{\mathcal{C}} \begin{bmatrix} V_x \\ V_y \end{bmatrix} \cdot \begin{bmatrix} dx \\ dy \end{bmatrix} = \int_{\mathcal{C}} \begin{bmatrix} \frac{\partial\phi}{\partial x} \\ \frac{\partial\phi}{\partial y} \end{bmatrix} \cdot \begin{bmatrix} dx \\ dy \end{bmatrix}$$

$$= \int_{\mathcal{C}} \frac{\partial\phi}{\partial x} dx + \frac{\partial\phi}{\partial y} dy = \int_{\mathcal{C}} d\phi = \phi(b) - \phi(a) \tag{3.31}$$

It is clear that the result of the line integral is always $\phi(b) - \phi(a)$ regardless of how $\mathcal{C}$ is chosen. Equation (3.31) represents the **fundamental theorem for line integrals**. For any closed path $\mathcal{C}$, where $b = a$, the work done is always zero. Due to this property, a gradient field is also referred to as a **conservative field**.

Additionally, if $\mathcal{C}$ corresponds to a contour line of ϕ, the work done is zero, as shown in Eq. (3.29). This implies that when moving along a constant potential curve, the force vector does no work, indicating energy conservation behavior.

For instance, when walking on a flat horizontal surface, the gravitational field does no work regardless of the distance traveled.

To summarize, any conservative field $\mathbf{V}$ has the following properties:

1. $\mathbf{V}$ is a gradient field, implying that a potential function ϕ can be found such that $V_x = \frac{\partial \phi}{\partial x}$ and $V_y = \frac{\partial \phi}{\partial y}$.

2. $\frac{\partial V_y}{\partial x} = \frac{\partial V_x}{\partial y}$. This implies that the component functions are differentiable. This relation can be derived from item 1 under the condition that ϕ is continuously differentiable:

$$
\begin{aligned}
\frac{\partial V_y}{\partial x} &= \frac{\partial}{\partial x}\frac{\partial \phi}{\partial y} = \frac{\partial^2 \phi}{\partial x \partial y} \\
\frac{\partial V_x}{\partial y} &= \frac{\partial}{\partial y}\frac{\partial \phi}{\partial x} = \frac{\partial^2 \phi}{\partial y \partial x} = \frac{\partial^2 \phi}{\partial x \partial y} = \frac{\partial V_y}{\partial x}
\end{aligned}
\tag{3.32}
$$

3. The work done around any closed path is zero: $\oint_C \mathbf{V} \cdot d\mathbf{r} = 0$. This is evident from Eq. (3.31) when $b = a$.
4. The work done depends only on the starting point a and the ending point b, not on the path chosen. This is described by Eq. (3.31).
5. The work done along any C that is a contour of ϕ is zero: $\int_C \mathbf{V} \cdot d\mathbf{r} = 0$.

Note that the constant in a potential function ϕ is immaterial in terms of computing the vector components because the component functions are obtained using its partial derivatives. Any additive constant changes only the contour lines of the potential function. Therefore, we need only to find the nonconstant part of the potential function ϕ if it exists for a given vector field.

Let us see some examples.

3.4.4 *Example: A nonconservative spin field*

This example shows that the spin field

$$
\mathbf{s} = -y\mathbf{i} + x\mathbf{j} = \begin{bmatrix} -y \\ x \end{bmatrix}
\tag{3.33}
$$

is nonconservative and the potential function cannot be found.

First, we know that the component functions are all polynomial and differentiable. We thus use the following snippet to perform the differentiations and show that item 2 cannot be satisfied:

```
1  # To begin with, check if item 2 is satisfied:
2  x, y = symbols('x, y', real=True)                    # define variables
3  X = sp.Matrix([x, y])
4
5  Vx = -y                                 # a spin field sz
6  Vy =  x
7  dVydx = Vy.diff(x)
8  dVxdy = Vx.diff(y)
9  display(Math(f" \\frac{{∂V_x}}{{∂y}}=\\frac{{∂V_y}}{{∂x}}?\;\; \
10                     {latex((dVxdy-dVydx).simplify()==0)}"))
```

$$\frac{\partial V_x}{\partial y} = \frac{\partial V_y}{\partial x}? \text{ False}$$

```
1  # Optional check:
2  curl_sz = sp.simplify(gr.curl_vf(Matrix([Vx, Vy]), X))
3  display(Math(f" \\text{{ Curl of s$_z$}}={latex(curl_sz)}"))
4
5  div_sz = sp.simplify(gr.div_vf(Matrix([Vx, Vy]), X))
6  display(Math(f" \\text{{ Divergence of s$_z$}}={latex(div_sz)}"))
```

Curl of $\mathbf{s}_z = 2$

Divergence of $\mathbf{s}_z = 0$

It is not a conservative field. It has a constant curl and zero divergence, which is typical for a spin field. However, let's see if we can find the potential function for this field:

```
1  #First, use ∂φ/∂x =-y:
2  C = sp.Function('C')(y)                   # define symbolic integral function
3  φ_C = integrate(-y, x) + C
4  φ_C                                       # possible potential function
```

$$-xy + C(y)$$

```
1  #Next, use ∂φ/∂y = x:
2  D = sp.Function('D')(x)                   # define symbolic integral function
3  φ_D = integrate(x, y) + D
4  φ_D                                       # possible potential function
```

$$xy + D(x)$$

Since $C(y)$ is only a function of y and $D(x)$ is only a function of x, it is not possible to find such $C(y)$ and $D(x)$ that make ϕ_C equal to ϕ_D. Thus, a potential function ϕ cannot be found. On the other hand, as we will see in the following chapter, a so-called stream function can be found for divergence-free fields, including spin fields.

Since the spin field is nonconservative, the work done by it will be path-dependent, as shown earlier using the two paths given in Fig. 3.8 and the elliptic helix path given in Fig. 3.9.

3.4.5 *Example: A conservative position field*

Since $\mathbf{r} = x\mathbf{i} + y\mathbf{j}$, to prove that $\mathbf{r}$ is a gradient field, we need to find a scalar potential function $\phi(x, y)$ such that

$$\frac{\partial \phi}{\partial x} = x; \quad \frac{\partial \phi}{\partial y} = y \tag{3.34}$$

The position field is also differentiable. The following snippet is used to perform the differentiations and show that item 2 is satisfied:

```
1  # To begin with, check if item 2 is satisfied:
2  Vx = x; Vy = y                              # position field
3  dVydx = Vy.diff(x)
4  dVxdy = Vx.diff(y)
5  display(Math(f" \\frac{{∂V_x}}{{∂y}}=\\frac{{∂V_y}}{{∂x}}?\;\; \
6                  {latex((dVxdy-dVydx).simplify()==0)}"))
```

$$\frac{\partial V_x}{\partial y} = \frac{\partial V_y}{\partial x}? \text{ True}$$

Next, we use the following code to find such a $\phi(x, y)$ using conditions given in Eq. (3.30) (or the conditions in item 1):

```
1  # First, use condition: x = ∂φ/∂x to find possible φ:
2  C = sp.Function('C')(y)              # define symbolic integral function
3  φ_C = integrate(Vx, (x)) + C         # add an unknown integral function
4  φ_C                                  # possible potential function
```

$$\frac{x^2}{2} + C(y)$$

We need to determine $C(y)$ that is a function of y:

```
1  # Next, use condition: y = ∂φ/∂y to find C(y)
2  φ_C.diff(y)
```

$$\frac{d}{dy} C(y)$$

Since $\frac{\partial C(y)}{\partial y} = y$, we have $C(y) = \frac{1}{2}y^2$. Here, we basically solved the differential equation $\frac{\partial C(y)}{\partial y} = y$ manually. Now, substituting $C(y)$ to the expression of ϕ_C, we obtain the following:

```
1  ϕ = ϕ_C.subs(C, y**2/2)
2  display(Math(f" \\text{{The potential function found = }}{latex(ϕ)}"))
```

The potential function found $= \dfrac{x^2}{2} + \dfrac{y^2}{2}$

Lastly, we check whether the potential function found satisfies the conditions in item 1:

```
1  print(f"Is x = ϕ.diff(x)? {Vx == ϕ.diff(x)}")
2  print(f"Is y = ϕ.diff(y)? {Vy == ϕ.diff(y)}")
```

```
Is x = ϕ.diff(x)? True
Is y = ϕ.diff(y)? True
```

Finally, the potential function for the position field (or radial field) can be expressed as

$$\phi(x,y) = \frac{1}{2}(x^2 + y^2) \tag{3.35}$$

It is continuously differentiable in the entire domain. The contours of the potential function are circles all centered at the origin. With such potentials, the position vector field radiates as shown in Fig. 3.2.

In conclusion, this position field is a gradient field and is conservative with circular potential contours:

```
1  # Optional check:
2  curl_r = sp.simplify(gr.curl_vf(Matrix([Vx, Vy]), X))
3  display(Math(f" \\text{{ Curl of r}}={latex(curl_r)}"))
4
5  div_r = sp.simplify(gr.div_vf(Matrix([Vx, Vy]), X))
6  display(Math(f" \\text{{ Divergence of r}}={latex(div_r)}"))
```

Curl of $\mathbf{r} = 0$

Divergence of $\mathbf{r} = 2$

It is curl-free and has a constant divergence, which is typical for a position field.

3.4.6 *Example: Radial field* **r**/r, *a gradient field*

Consider a radial field defined as

$$\mathbf{r}_1 = \begin{bmatrix} \frac{x}{r} \\ \frac{y}{r} \end{bmatrix} \tag{3.36}$$

where $r = \sqrt{x^2 + y^2}$. This radial field has unit amplitude.

Show that this radial field is a gradient field.

Solution: The procedure is largely the same as the previous example:

```
1  # To begin with, check if item 2 is satisfied:
2  x, y = symbols('x, y', real=True)              # define variables
3  X = sp.Matrix([x, y])
4
5  Vx = x/sp.sqrt(x**2+y**2)                       # a position field r1
6  Vy = y/sp.sqrt(x**2+y**2)
7  dVydx = Vy.diff(x)
8  dVxdy = Vx.diff(y)
9  display(Math(f" \\frac{{∂V_x}}{{∂y}}=\\frac{{∂V_y}}{{∂x}}?\;\; \
10               {latex((dVxdy-dVydx).simplify()==0)}"))
```

$$\frac{\partial V_x}{\partial y} = \frac{\partial V_y}{\partial x}? \quad \text{True}$$

We use the following code to find the potential function ϕ for this radial field. The procedure is largely the same as the previous example:

```
1  # First, use condition: r_x = ∂φ/∂x to find a possible φ:
2
3  C = sp.Function('C')(y)          # define symbolic integral function
4  φC = integrate(Vx, x) + C                # add unknown integral func.
5  φC                                       # possible potential function
```

$$\sqrt{x^2 + y^2} + C(y)$$

```
1  # Second: use condition: r_y = ∂φ/∂y to find possible φ:
2
3  D = sp.Function('D')(x)          # define symbolic integral function
4  φD = integrate(Vy, y) + D                # add unknown integral func.
5  φD                                       # possible potential function
```

$$\sqrt{x^2 + y^2} + D(x)$$

Since $C(x)$ is a function of only x, and $D(y)$ is a function of only y, to find a common ϕ, setting both $C(x)$ and $D(y)$ to zero, we obtain

$$\phi(x,y) = \sqrt{x^2 + y^2} \qquad (3.37)$$

We have successfully found the potential $\phi(x,y)$, and hence this radial field is a gradient field and conservative. The contours of the potential function are circles (slices of a cone) with the origin as the center. With such potentials, the radial field radiates in a way similar to that shown in Fig. 3.2, with equal spatial amplitude.

Finally, let us check whether condition item 2 is satisfied:

```
1  φ = sp.sqrt(x**2 + y**2)
2  Vy = φ.diff(y); Vx = φ.diff(x)
3
4  print(f"Is item 2 true: {(Vy.diff(x)-Vx.diff(y)).simplify()==0}")
5  display(Math(f" \\frac{{∂φ}}{{∂x}}={latex(φ.diff(x))}; \\;\\; \
6                 \\frac{{∂φ}}{{∂y}}={latex(φ.diff(y))}"))
```

Is item 2 true: True

$$\frac{\partial \phi}{\partial x} = \frac{x}{\sqrt{x^2 + y^2}}; \quad \frac{\partial \phi}{\partial y} = \frac{y}{\sqrt{x^2 + y^2}}$$

```
1  # Optional check:
2  curl_r1 = sp.simplify(gr.curl_vf(Matrix([Vx, Vy]), X))
3  display(Math(f" \\text{{ Curl of r$_1$}}={latex(curl_r1)}"))
4
5  div_r1 = sp.simplify(gr.div_vf(Matrix([Vx, Vy]), X))
6  display(Math(f" \\text{{ Divergence of r$_1$}}={latex(div_r1)}"))
```

Curl of $\mathbf{r}_1 = 0$

Divergence of $\mathbf{r}_1 = \dfrac{1}{\sqrt{x^2 + y^2}}$

It is curl-free and has a divergence that decays with r.

3.4.7 *Example: Radial field* $\mathbf{r}/r^2$, *a gradient field*

Consider a radial field defined as

$$\mathbf{r}_2 = \begin{bmatrix} \frac{x}{r^2} \\ \frac{y}{r^2} \end{bmatrix} \qquad (3.38)$$

where $r = \sqrt{x^2 + y^2}$.

Show that this radial field is a gradient field.

Solution: It is obvious that the field has a singularity at the origin. It is not differentiable there. Let us check item 2 for the field, excluding the origin:

```
1  # To begin with, check if item 2 is satisfied:
2  x, y = symbols('x, y', real=True)              # define variables
3  X = sp.Matrix([x, y])
4
5  Vx = x/(x**2+y**2)                              # a position field r2
6  Vy = y/(x**2+y**2)
7  dVydx = Vy.diff(x)
8  dVxdy = Vx.diff(y)
9  display(Math(f" \\frac{{∂V_x}}{{∂y}}=\\frac{{∂V_y}}{{∂x}}?\;\;  \
10                {latex((dVxdy-dVydx).simplify()==0)}"))
```

$$\frac{\partial V_x}{\partial y} = \frac{\partial V_y}{\partial x}? \ \text{True}$$

We use the following code to find the potential function for this radial field:

```
1  # First, use condition: r2_x = ∂φ/∂x to find a possible φ:
2
3  C = Function('C')(x)
4  r2x = x/(x**2+y**2)
5  φC = integrate(r2x, x) + C
6  φC
```

$$C(x) + \frac{\log\left(x^2 + y^2\right)}{2}$$

```
1  # Second, use condition: r2_y = ∂φ/∂y to find a possible φ:
2
3  D = Function('D')(y)
4  r2y = y/(x**2+y**2)
5  φD = integrate(r2y, y) + D
6  φD
```

$$D(y) + \frac{\log\left(x^2 + y^2\right)}{2}$$

Setting both $C(x)$ and $D(y)$ to zero, we obtain

$$\phi(x, y) = \frac{1}{2}\log(x^2 + y^2) \tag{3.39}$$

We have found the potential $\phi(x, y)$, and hence the radial field is a gradient field. The contours of the potential function are circles centered at the origin, and the radial vector field radiates similar to that shown in Fig. 3.2, with a fast-decaying spatial amplitude.

Let us check whether condition item 2 is satisfied:

```
1  ϕ = sp.log(x**2 + y**2)/2
2  Vy = ϕ.diff(y); Vx = ϕ.diff(x)
3
4  print(f"Is item 2 true: {(Vy.diff(x)-Vx.diff(y)).simplify()==0}")
5  display(Math(f" \\frac{{∂ϕ}}{{∂x}}={latex(ϕ.diff(x))}; \\;\\; \
6                \\frac{{∂ϕ}}{{∂y}}={latex(ϕ.diff(y))}"))
```

Is item 2 true: True

$$\frac{\partial \phi}{\partial x} = \frac{x}{x^2 + y^2}; \quad \frac{\partial \phi}{\partial y} = \frac{y}{x^2 + y^2}$$

```
1  # Optional check:
2  curl_r2 = sp.simplify(gr.curl_vf(Matrix([Vx, Vy]), X))
3  display(Math(f" \\text{{ Curl of r$_2$}}={latex(curl_r2)}"))
4
5  div_r2 = sp.simplify(gr.div_vf(Matrix([Vx, Vy]), X))
6  display(Math(f" \\text{{ Divergence of r$_2$}}={latex(div_r2)}"))
```

Curl of $\mathbf{r}_2 = 0$
Divergence of $\mathbf{r}_2 = 0$

This field is curl-free and also divergence-free, which is a suspicious behavior or the field must be very special. The reason is the singularity at $(0,0)$, at which the field is not differentiable. We will discuss in more detail in Section 4.4.7 of Chapter 4.

3.4.8 *Example: Spin fields $/r^2$, a gradient field*

Consider a spin field defined as

$$\mathbf{s}_{r2} = \begin{bmatrix} -\frac{y}{r^2} \\ \frac{x}{r^2} \end{bmatrix} \tag{3.40}$$

where $r = \sqrt{x^2 + y^2}$. This is the spin field $\mathbf{s}_z$ (which is not a gradient field) divided by r^2.

Show that this spin field is a gradient field.

Solution: It is obvious that the field has a singularity at the origin. It is not differentiable there. Let us check item 2 for the field, excluding the origin:

```
1  # To begin with, check if item 2 is satisfied:
2
3  Vx = -y/(x**2+y**2)                              # a position field sr2
4  Vy =  x/(x**2+y**2)
5  dVydx = Vy.diff(x).simplify()
6  dVxdy = Vx.diff(y).simplify()
7  display(Math(f" \\frac{{∂V_x}}{{∂y}}=\\frac{{∂V_y}}{{∂x}}?\;\; \
8                     {latex((dVxdy-dVydx).simplify()==0)}"))
```

$$\frac{\partial V_x}{\partial y} = \frac{\partial V_y}{\partial x}? \ \text{True}$$

```
1  # Optional check:
2  curl_s2 = sp.simplify(gr.curl_vf(Matrix([Vx, Vy]), X))
3  display(Math(f" \\text{{ Curl of s$_2$}}={latex(curl_s2)}"))
4
5  div_s2 = sp.simplify(gr.div_vf(Matrix([Vx, Vy]), X))
6  display(Math(f" \\text{{ Divergence of s$_2$}}={latex(div_s2)}"))
```

Curl of $s_2 = 0$

Divergence of $s_2 = 0$

The spin field is also quite special. Both its curl and divergence are zero.

We use the following snippet to find the potential function for this field. The procedure is the same as in the previous example:

```
1  # First, use condition: sr2_x = ∂φ/∂x to find a possible φ:
2
3  x, y = symbols('x, y', real=True)
4  s2x = -y/(x**2+y**2)
5  φC = integrate(s2x, x) + C
6  φC
```

$$C(x) - \text{atan}\left(\frac{x}{y}\right)$$

```
1  # Second, use condition: sr2_y = ∂φ/∂y to find a possible φ:
2
3  s2y = x/(x**2+y**2)
4  φD = integrate(s2y, y) + D
5  φD
```

$$D(y) + \text{atan}\left(\frac{y}{x}\right)$$

Setting both $C(x)$ and $D(y)$ to zero, we obtain

$$\phi(x, y) = \arctan\left(\frac{y}{x}\right) \tag{3.41}$$

The potential function is in fact the polar angle $\theta = \arctan(\frac{y}{x})$, which drives this field. Let us check the item 2 condition using the field generated by this potential:

```
1  ϕ = sp.atan(y/x)
2  Fx = ϕ.diff(x).simplify(); Fy = ϕ.diff(y).simplify()
3  print(f"Is item 2 true: {(Fy.diff(x)-Fx.diff(y)).simplify()==0}")
4  display(Math(f" \\frac{{∂ϕ}}{{∂x}}={latex(Fx)}; \\;\\; \
5                \\frac{{∂ϕ}}{{∂y}}={latex(Fy)}"))
```

Is item 2 true: True

$$\frac{\partial \phi}{\partial x} = -\frac{y}{x^2 + y^2}; \quad \frac{\partial \phi}{\partial y} = \frac{x}{x^2 + y^2}$$

Although the spin field s_z is not a gradient field, when it is divided by r^2, it becomes one. It has a singularity at the origin, creating a strong radiation from the center. This converts it into a radial field, and hence it becomes a gradient field. Let us plot it to take a look:

```
1   from matplotlib import cm                      # needed for the plot
2   plt.rcParams.update({'font.size': 8})
3   plt.rcParams['figure.dpi'] = 500
4   a, b = 1., 1.                          # domain [-a, a]×[-b, b] for plots
5
6   fx = lambda x,y: -y/(x**2+y**2);         # need to use numpy function
7   fy = lambda x,y:  x/(x**2+y**2)
8
9   plt = plotField(a, b, fx, fy, 's_r2_Field',scale=30) # plot name as str
10  #plt.show()
```

The vector field s_{r2} is radial and, in fact, conservative (excluding the singular point at the center). This behavior is attributed to its singularity, as shown in Fig. 3.11:

> Singularities can often alter the characteristics of a field. Therefore, caution is necessary when analyzing fields with singularities.

The field will be further examined in the following chapter when studying Green's circulation theorem.

3.4.9 Example: Potential function of the gravity field

We use the following snippet to first confirm item 2 and then use item 1 to find the potential function $\phi(x, y)$ for the gravity field defined in Eq. (3.18):

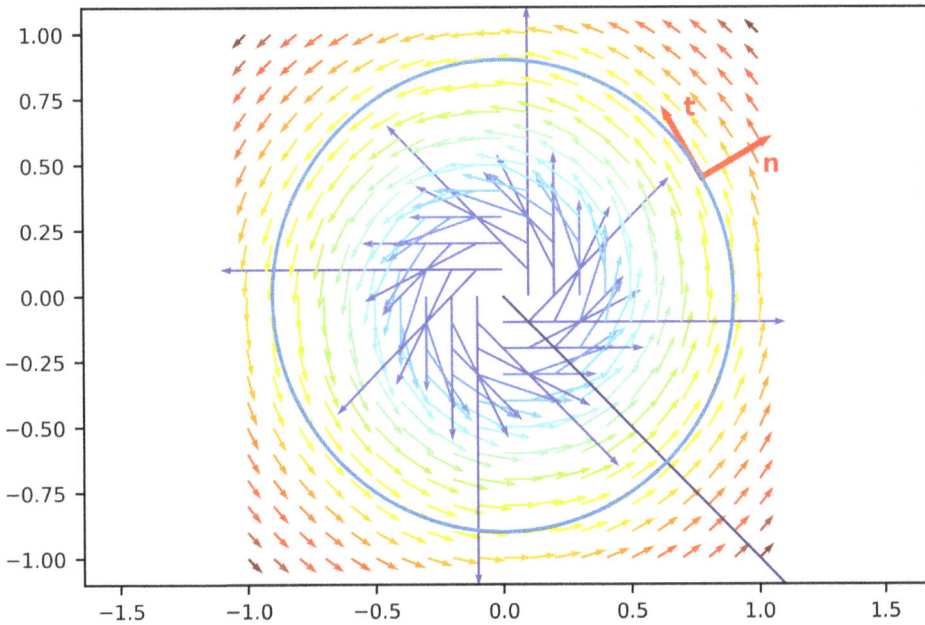

Figure 3.11. Vector field $\mathbf{s}_{r2}$, which is radial due to the singularity at the origin. The arrow (black) at the singular point becomes infinitely long.

```
1  x, y, t, a, b = symbols('x, y, t, a, b', real=True)  # define variables
2  X = sp.Matrix([x, y])
3
4  m, g, H = symbols('m, g, H', positive=True)
5  Gx, Gy = symbols('G_x, G_y', real=True)
6
7  Gx = 0*y                          # force in the x-direction
8  Gy = -m*g               # force in the y-direction, (kg m/s^2 = N)
9  Gv = sp.Matrix([Gx, Gy])
10
11 dGydx = Gy.diff(x).simplify()
12 dGxdy = Gx.diff(y).simplify()
13 # Check item 2:
14 display(Math(f" \\frac{{∂G_x}}{{∂y}}=\\frac{{∂G_y}}{{∂x}}?\;\; \
15                 {latex((dGxdy-dGydx).simplify()==0)}"))
```

$$\frac{\partial G_x}{\partial y} = \frac{\partial G_y}{\partial x}? \;\; \text{True}$$

Therefore, it is a conservative field. Let us find the potential function ϕ for the gravity field:

```
1  # Use condition: Gx = ∂φ/∂x to find possible φ:
2
3  C = sp.Function('C')(y)              # define symbolic integral function
4  φC = integrate(Gx, (x)) + C          # add an unknown integral function
5  φC                                   # possible potential function
```

$C(y)$

This means that the potential function can only be a function of y. Next:

```
1  # Use condition: Gy = ∂φ/∂y to solve for C(y):
2
3  sln_C = sp.dsolve(Gy - φC.diff(y), C) # use the differential eq. solver
4  sln_C.args                            # solution of integral function C(y)
```

$(C(y),\ C_1 - gmy)$

This time, we used the SymPy built-in sp.dsolve() to find $C(y)$. The second term in the tuple is the solution for $C(y)$. We can now put the solution of $C(y)$ back to ϕ_C:

```
1  φ = φC.subs(C,sln_C.args[1])
2  φ.args
```

$(C_1,\ -gmy)$

The nonconstant part of the solution is in the second term in the tuple. Finally, we can perform the following:

```
1  # Set C_1 to zero to find the final solution for potential function φ:
2
3  φ = φ.subs(φ.args[0],0)
4  display(Math(f" \\text{{The potential function φ = }}{latex(φ)}"))
```

The potential function $\phi = -gmy$

```
1  # Finally let's Check:
2
3  print(f"Is Fx = φ.diff(x)? {Gx == φ.diff(x)}")
4  print(f"Is Fy = φ.diff(y)? {Gy == φ.diff(y)}")
```

```
Is Fx = φ.diff(x)? True
Is Fy = φ.diff(y)? True
```

We successfully found the potential function $\varphi = -mgy$ with the help of SymPy. It depends on the coordinate y as expected. The larger the y, the more negative the potential due to gravity. The negative sign is because the gravitational field is directed downward.

In conclusion, the gravitational field we experience is a gradient field and is conservative. It does no work along any closed loop, and the work done is path-independent, determined only by the locations of the start and end points of the path.

The example problems given above are relatively simple, and the codes provided detail the procedure for finding a potential function for a given field and how to find it manually.

In the last example, we used `sp.dsolve()`, a differential equation solver. While we have not discussed this in detail, it can be used here as a tool. Although the process could be done without it, the approach would be less straightforward and more difficult to automate, as shown in the previous examples. By using `sp.dsolve()`, we can now write a Python function to find the potential function automatically, provided a potential function exists for the given 2D vector field.

3.4.10 *Code for finding the potential function of a vector field*

```
1  def find_potential_f2D(Vx, Vy, X):
2      '''
3          Find the potential function φ for a given two vector field
4          components in 2D. The vector field must be a gradient field.
5          return: φ, potential function.
6      '''
7      x, y = X
8      # Use condition: ∂φ/∂x = Vx
9      C = sp.Function('C')(y)          # define symbolic integral functions
10     φC = sp.integrate(Vx, (x)) + C      # φ with C(y); used Vx = ∂φ/∂x
11
12     # use condition: Vy = ∂φ/∂y
13     eqn = φC.diff(y)- Vy                 # create an equation to solve
14     sln_C = sp.dsolve(eqn, C)             # solve the eqn for C(y)
15
16     φC = φC.subs(C, sln_C.args[1])    # put solution of C(y) back to φC
17     φ  = φC.subs(φC.args[0],0).simplify() # set integral constant to 0
18     gr.printx('φ')
19
20     # Check all conditions:
21     gr.printx('(φ.diff(x)-Vx).simplify()') # check condition: Vx=∂φ/∂x
22     gr.printx('(φ.diff(y)-Vy).simplify()') # check condition: Vy=∂φ/∂y
23
24     return φ
```

Let us use find_potential_f2D() to find the potential function for the gravity field:

```
1  φ = find_potential_f2D(Gx, Gy, X)
2  gr.printM(φ, ' The potential function φ found is: ')
```

```
φ = -g*m*y
(φ.diff(x)-Vx).simplify() = 0
(φ.diff(y)-Vy).simplify() = 0
 The potential function φ found is:
```

$$-gmy$$

We obtained the same result as earlier. We also test it for field s_{r2} studied previously:

```
1  φ = find_potential_f2D(s2x, s2y, X)
2  gr.printM(φ, ' The potential function φ found is: ')
```

```
φ = -atan(x/y)
(φ.diff(x)-Vx).simplify() = 0
(φ.diff(y)-Vy).simplify() = 0
 The potential function φ found is:
```

$$-\operatorname{atan}\left(\frac{x}{y}\right)$$

This result matches the one we obtained earlier. Readers may test other cases.

Care must be taken when using **find_potential_f2D()**. If the field is nonconservative, the code may still produce a function, but the output of the tests will fail. For example, for the spin field that was proven to be nonconservative, the results are as follows:

```
1  φ = find_potential_f2D(-y, x, X)
```

```
φ = x*y
(φ.diff(x)-Vx).simplify() = 2*y
(φ.diff(y)-Vy).simplify() = 0
```

As observed, the function found failed the test, indicating it is not a true potential function.

Note also that **find_potential_f2D()** integrates over x first and then y, which may not always be the optimal sequence. For some fields, the correct potential function might be found by reversing the integration order. This can be achieved by swapping the components V_x and V_y without modifying the code.

3.4.11 *Case study: Constructing gradient fields*

Assume the component functions of 2D fields have the following character-istics:

- **Case 1:** One of the components is constant.
- **Case 2:** Both component functions are linear.

Determine possible formulas for the gradient fields and potential functions for these two cases.

Solution: Consider Case 1. Assume the x-component of the vector field is constant, denoted by the symbolic variable c, and let the y-component be an unknown function represented by the symbolic function F_y. Using find_potential_f2D(), we can find the potential function. The following snippet performs this task:

```
1  x, y, c = symbols('x, y, c', real=True)           # define variables
2  Fy = sp.Function('F_y')(y)                 # define the unknown function
3  Fx = sp.Function('F_x')(x)
4
5  Vx = c; Vy = Fy                            # define vector field
6  ϕ = find_potential_f2D(c, Fy, X)
7  display(Math(f" \\text{{The potential function ϕ}}={latex(ϕ)}"))
```

```
ϕ = c*x + Integral(F_y(y), y)
(ϕ.diff(x)-Vx).simplify() = 0
(ϕ.diff(y)-Vy).simplify() = 0
```

The potential function $\phi = cx + \displaystyle\int F_y(y)\,\mathrm{d}y$

The potential function for the gradient fields for Case 1 is

$$\phi(x, y) = cx + \int F_y(y)\,\mathrm{d}y$$

$$\text{or}\quad \phi(x, y) = cy + \int F_x(x)\,\mathrm{d}x$$

(3.42)

where c is an arbitrary constant. The corresponding vector field functions are

$$\mathbf{F}(x, y) = \begin{bmatrix} c \\ F_y \end{bmatrix} \quad \text{or} \quad \mathbf{F}(x, y) = \begin{bmatrix} F_x \\ c \end{bmatrix}$$

(3.43)

```
1  # Check:
2  Vx = Fx; Vy = c                                    # define vector field
3  φ = find_potential_f2D(Vx, Vy, X)                  # find its potential
4  display(Math(f" \\text{{The potential function φ}}={latex(φ)}"))
```

```
φ = c*y + Integral(F_x(x), x)
(φ.diff(x)-Vx).simplify() = 0
(φ.diff(y)-Vy).simplify() = 0
```

The potential function $\phi = cy + \int F_x(x)\,\mathrm{d}x$

We draw the following **conclusion**:

> If either component function is constant, the field is a gradient field, and the potential function will be cx plus the antiderivative of F_y, or cy plus the antiderivative of F_x, provided the antiderivatives exist.

Solution: Consider Case 2. Assume the two components of the vector field are independent, general linear functions. We can define these functions symbolically and use **find_potential_f2D()** to determine the potential functions. The following snippet accomplishes this:

```
 1  x, y = symbols('x, y', real=True)                       # define variables
 2  X = sp.Matrix([x, y])
 3  a1, b1, c1 = symbols('a1, b1, c1', real=True)           # define constants
 4  a2, b2, c2 = symbols('a2, b2, c2', real=True)
 5  Fx = sp.Function('F_x')(x,y)                    # define the nuknown functions
 6  Fy = sp.Function('F_y')(x,y)
 7
 8  Fx = a1*x + b1*y + c1
 9  Fy = a2*x + b2*y + c2
10  φ = find_potential_f2D(Fx, Fy, X)
11  display(Math(f" \\text{{The potential function φ}}={latex(φ)}"))
```

```
φ = a1*x**2/2 + a2*x*y + b2*y**2/2 + c1*x + c2*y
(φ.diff(x)-Vx).simplify() = y*(a2 - b1)
(φ.diff(y)-Vy).simplify() = 0
```

The potential function $\phi = \dfrac{a_1 x^2}{2} + a_2 xy + \dfrac{b_2 y^2}{2} + c_1 x + c_2 y$

The results show that if $a_2 = b_1$, the function found is a potential function. Due to the symmetry property, we know that if $b_2 = a_1$, the corresponding function must also be a potential function. Thus, we have

$$\phi(x, y) = \frac{a_1 x^2}{2} + \frac{b_2 y^2}{2} + c_2 y + x (b_1 y + c_1)$$

or   (3.44)

$$\phi(x, y) = \frac{a_2 x^2}{2} + \frac{b_1 y^2}{2} + c_1 y + x (a_1 y + c_2)$$

The corresponding vector field functions are

$$\mathbf{F}(x, y) = \begin{bmatrix} a_1 x + b_1 y + c_1 \\ b_1 x + b_2 y + c_2 \end{bmatrix} \quad \text{or} \quad \mathbf{F}(x, y) = \begin{bmatrix} a_2 x + a_1 y + c_2 \\ a_1 x + b_1 y + c_1 \end{bmatrix} \quad (3.45)$$

where a_1, b_1, c_1 and a_2, b_2, c_2 are arbitrary constants. Let us confirm all of this using the following snippets:

```
1  # Check:
2  Vx = a1*x + b1*y + c1                        # define vector field
3  Vy = b1*x + b2*y + c2                        # set a2 = b1
4  ϕ = find_potential_f2D(Vx, Vy, X)           # find its potential
5  display(Math(f" \\text{{The potential function ϕ}}={latex(ϕ)}"))
```

```
ϕ = a1*x**2/2 + b2*y**2/2 + c2*y + x*(b1*y + c1)
(ϕ.diff(x)-Vx).simplify() = 0
(ϕ.diff(y)-Vy).simplify() = 0
```

The potential function $\phi = \dfrac{a_1 x^2}{2} + \dfrac{b_2 y^2}{2} + c_2 y + x (b_1 y + c_1)$

```
1  # Check:
2  Vx = a2*x + a1*y + c2                        # swap Vx with Vy, and set b2 = a1
3  Vy = a1*x + b1*y + c1                        # find its potential
4  ϕ = find_potential_f2D(Vx, Vy, X)
5  display(Math(f" \\text{{The potential function ϕ}}={latex(ϕ)}"))
```

```
ϕ = a2*x**2/2 + b1*y**2/2 + c1*y + x*(a1*y + c2)
(ϕ.diff(x)-Vx).simplify() = 0
(ϕ.diff(y)-Vy).simplify() = 0
```

The potential function $\phi = \dfrac{a_2 x^2}{2} + \dfrac{b_1 y^2}{2} + c_1 y + x (a_1 y + c_2)$

We draw the following conclusion:

> If the component functions are linear, the vector field is a gradient vector field if the coefficient of x in the second component function equals the coefficient of y in the first component function.

3.4.12 *A list of often used gradient fields*

The following is a list of often used gradient fields provided by ChatGPT:

Field	**Gradient Field**
Electrostatic field (electric field)	$\mathbf{E} = -\nabla\phi(\mathbf{r})$
Gravitational field	$\mathbf{g} = -\nabla V(\mathbf{r})$
Temperature field in heat conduction	$\mathbf{q} = -k\nabla T(\mathbf{r}, t)$
Pressure gradient in fluid dynamics	$\nabla P(\mathbf{r}, t)$
Concentration gradient in diffusion	$\mathbf{J} = -D\nabla C(\mathbf{r}, t)$
Potential energy gradient in mechanics	$\mathbf{F} = -\nabla U(\mathbf{r})$
Magnetic scalar potential	$\mathbf{B} = -\nabla\phi_m(\mathbf{r})$
Chemical potential in thermodynamics	$\mathbf{F}_{\text{diffusion}} = -\nabla\mu(\mathbf{r}, t)$
Height or elevation gradient in geography	$\nabla h(\mathbf{r})$
Wavefront gradient in optics (Eikonal equation)	$\mathbf{k} = \nabla S(\mathbf{r})$

$$(3.46)$$

3.5 Conservation of total energy

When a gravitational force field does work on a mass (m), a change in the velocity of the mass results in a change in its kinetic energy. Based on Newton's law, the work done by the force can be evaluated as

$$\int_a^b \mathbf{F} \cdot d\mathbf{R} = \int m\frac{d\mathbf{v}}{dt} \cdot \mathbf{v}dt = \frac{1}{2}m[\mathbf{v} \cdot \mathbf{v}]\Big|_a^b$$

$$= \frac{1}{2}m\|\mathbf{v}(b)\|^2 - \frac{1}{2}m\|\mathbf{v}(a)\|^2 \tag{3.47}$$

On the other hand, the work done can be found as using the potential function of the gravity field:

$$\int_a^b \mathbf{F} \cdot d\mathbf{R} = -\int_a^b d\phi = \phi(a) - \phi(b) \tag{3.48}$$

where ϕ is the potential function. The minus sign is due to the fact that the work done by gravity reduces the potential.

Since Eqs. (3.47) and (3.48) must be equal, we obtain

$$\frac{1}{2}m\|\mathbf{v}(b)\|^2 - \frac{1}{2}m\|\mathbf{v}(a)\|^2 = \phi(a) - \phi(b) \tag{3.49}$$

Rearranging gives

$$\underbrace{\frac{1}{2}m\|\mathbf{v}(b)\|^2 + \phi(b)}_{\text{total energy at } b} = \underbrace{\frac{1}{2}m\|\mathbf{v}(a)\|^2 + \phi(a)}_{\text{total energy at } a} \tag{3.50}$$

where the total energy is the sum of the kinetic energy $\frac{1}{2}m\|\mathbf{v}(\cdot)\|^2$ and the potential energy $\phi(\cdot)$). Equation (3.50) states that the **total energy is conserved**.

3.6 Flux of a vector field across a curve

We know how to measure the work done by a force field along a curve, as illustrated in Fig. 3.12(a). This time, we will measure the flux of a vector field, such as a velocity field, across a curve. The flux is measured in a direction rotated 90°, as shown in Fig. 3.12(b).

Consider a velocity field for a flowing medium. The amount of flow, called flux q, across curve $\mathcal{C}$ should be the integral of the dot product:

$$q = \int_{\mathcal{C}} \mathbf{V} \cdot \mathbf{n} \, ds \tag{3.51}$$

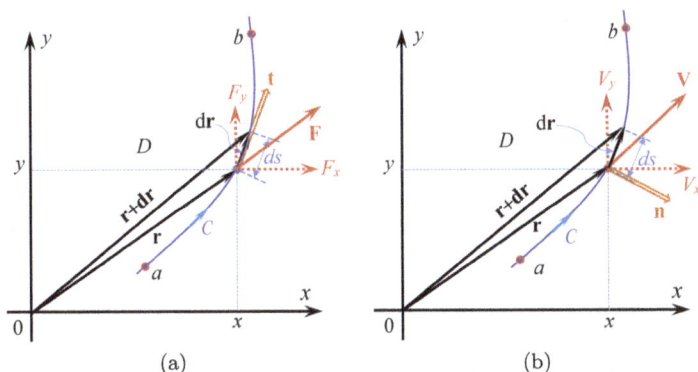

Figure 3.12. Measures of the effects of vector fields on a curve $\mathcal{C}$ at point (x, y) in a 2D plane: (a) the work done by a force field $\mathbf{F}$ along a curve $\mathcal{C}$, which involves the dot product $\mathbf{F} \cdot \mathbf{t} ds$ and (b) the flux of a velocity field $\mathbf{F}$ across a curve $\mathcal{C}$, which involves the dot product $\mathbf{V} \cdot \mathbf{n} ds$.

where **n** is the outward normal unit vector at point (x, y) on C and ds is the differential length along C at point (x, y). Equation (3.51) states the following:

1. If **V** at (x, y) is perpendicular to **n**, nothing will flow across C, so the flux across C is zero.
2. If **V** is in the same direction as **n**, the flux is $|\mathbf{V}|ds$ (out of the domain $\mathcal{D}$) in the direction of **n**.
3. If **V** is in the opposite direction to **n**, the flux is $-|\mathbf{V}|ds$ (into the domain $\mathcal{D}$) in the opposite direction to **n**.
4. Otherwise, only the component of **V** in the direction of **n** contributes to the flux across C. Thus, the dot product given in Eq. (3.51) produces a scalar q. If q is positive, the medium flows out; otherwise, it flows in.

3.6.1 *Normal vector on a curve*

As shown in Fig. 3.12(b), the normal vector **n** is a $90°$ (or clockwise) rotated tangent vector **t** (or $d\mathbf{r}/ds$). Therefore, **n** can be obtained using a rotation matrix **R**, as given in Eq. (2.16):

$$\mathbf{R} = \begin{bmatrix} \cos(-90°) & -\sin(-90°) \\ \sin(-90°) & \cos(-90°) \end{bmatrix} = \begin{bmatrix} 0 & 1 \\ -1 & 0 \end{bmatrix} \tag{3.52}$$

Thus, normal vector **n** is obtained using

$$\mathbf{n}ds = \mathbf{R}(d\mathbf{r}) = \begin{bmatrix} 0 & 1 \\ -1 & 0 \end{bmatrix} \begin{bmatrix} dx \\ dy \end{bmatrix} = \begin{bmatrix} dy \\ -dx \end{bmatrix} \tag{3.53}$$

Note here that the rotation causes dx and dy swapping places with a sign change in dx.

Next, using Eq. (3.53), Eq. (3.51) becomes

$$q = \int_C \mathbf{V} \cdot \mathbf{n}ds = \int_C V_x dy - V_y dx \tag{3.54}$$

It measures the total flux q out of the curve C of the domain $\mathcal{D}$. The flux here is a scalar, which differs from the concept of flux in transport problems, where the flux can be a gradient of a scalar function and thus a vector.

Let us see some examples.

3.6.2 *Example: Flux of the gravity field across curves*

Consider the gravity field studied in the example given in Section 3.2.3 and the same curve shown in Fig. 3.8. The vector field is given by Eq. (3.18), and all variables defined there can be used (you may need to revisit the example in Section 3.2.3 first and then execute the following snippets). The flux of the gravity field can be computed using the following snippets. First, consider the vertical curve C_1:

```
1  # Curve C1:
2  x_ = 0*t                                    # x_: x given in t
3  y_ = t                                      # y_: y given in t
4  dic = {x:x_, y:y_}                     # for later substitution
5  dnds = sp.Matrix([y_.diff(t), -x_.diff(t)])    # normal vector on C
6
7  # Use the components formula:
8  q_C1 = integrate(Gx.subs(dic),(t, H, 0))     # only need the y-component
9  display(Math(f" \\text{{Flux of the gravity force across C1, \
10                       via components formula = }} {latex(q_C1)} "))
11
12 # Use the dot-product formula:
13 q_C1 = integrate((Gv.T@dnds).subs(dic), (t, H, 0))[0]
14 display(Math(f" \\text{{Flux of the gravity force across C1, \
15                       via dot-product formula = }} {latex(q_C1)} "))
```

Flux of the gravity force across C_1 via components formula $= 0$

Flux of the gravity force across C_1 via dot-product formula $= 0$

We obtained zero as expected because the gravity field is vertical and curve C_1 is also vertical. It is not possible to have any flux across this curve.

Next, consider the curve C_2, which is a semi-circle on the right side of the upper plane:

```
1  R, y0 = symbols('R, y0', positive=True)
2  θ = symbols('θ ', real=True)
3  H = 100; R = H/2  #(m)
4  x_ = R*sp.cos(θ)                          # Cartesian-polar mapping
5  y_ = y0 + R*sp.sin(θ)
6
7  dnds = sp.Matrix([y_.diff(θ), -x_.diff(θ)])
8  q_C2 = sp.integrate((Gv.T@dnds).subs(dic), (θ, sp.pi/2, -sp.pi/2))[0]
9  display(Math(f" \\text{{Flux of the gravity force across C2,\
10                       via dot-product formula = }} {latex(q_C2)}"))
```

Flux of the gravity force across C_2 via dot-product formula $= 0$

The flux is still zero. The vertical gravity field flows into the upper part of the curve and out of the lower part of the curve (see Fig. 3.8). Thus, the net flux becomes zero.

If the semi-circle curve includes only the upper half (one quarter of the circle) or the lower half, we obtain the following:

```
1  # Use upper quarter-circle curve: θ from 0 to pi/2:
2  q_C2 = sp.integrate((Gv.T@dnds).subs(dic), (θ, 0, sp.pi/2))[0]
3  display(Math(f" \\text{{Flux of the gravity force across upper C2,\
4                          via dot-product formula = }} {latex(q_C2)}"))
5
6  # Use the quarter semi-circle curve: θ from -pi/2 to 0:
7  q_C2 = sp.integrate((Gv.T@dnds).subs(dic), (θ, -sp.pi/2, 0))[0]
8  display(Math(f" \\text{{Flux of the gravity force across lower C2,\
9                          via dot-product formula = }} {latex(q_C2)}"))
```

Flux of the gravity force across upper C_2 via dot-product formula $= -50.0\,\mathrm{gm}$
Flux of the gravity force across lower C_2 via dot-product formula $= 50.0\,\mathrm{gm}$

This time, we get a nonzero value, which is negative for the upper quarter circle because the gravity field is downward, and the flux is into the domain bounded by curves C_1 and C_2. The flux is positive for the lower quarter circle, where the gravity field's flux is out of the domain.

3.6.3 *Example: Flux of a spin field across curves*

Consider the same spin field defined in Eq. (3.23). We would like to compute the flux of this field, considering the same curves given in Fig. 3.8.

Solution: We use Eq. (3.54) and the same data from the example in Section 3.2.4. First, for curve C_1, we use the following snippet to compute the flux:

```
1  Fx = -y                                    # force in the x-direction
2  Fy =  x                                    # force in the y-direction
3  Fv = sp.Matrix([Fx, Fy])                         # force vector
4
5  x_ = 0*t                                   # use parameter t
6  y_ = t                                          # y = t
7  dic = {x:x_, y:y_, R:50}
8  dnds = sp.Matrix([y_.diff(t), -x_.diff(t)])     # normal vector on C
9
10 # use the two components:
11 q_C1 = integrate(Fx.subs(dic),(y,H,0))-integrate(Fy.subs(dic),(x,0,0))
12 q_C1 = q_C1.subs(t, y)                     # substitute t back with y
13 display(Math(f" \\text{{Flux of the spin field across C$_1$,\
14                         via component formula = }} {latex(q_C1)}"))
15 # use the dot-product:
16 q_C1 = integrate((Fv.T@dnds).subs(dic), (y, H, 0))[0]
17 q_C1 = q_C1.subs(t, y)                     # substitute t back with y
18 display(Math(f" \\text{{Flux of the spin field across C$_1$,\
19                         via dot-product = }} {latex(q_C1)}"))
```

Flux of the spin field across C_1 via component formula $= 100y$

Flux of the spin field across C_1 via dot product $= 100y$

We obtained the same positive value using both formulas. The results make sense because the spin field does flow across curve C_1, in its normal direction, and out of the domain bounded by curves C_1 and C_2.

Let us change the center of the semi-circle C_2 to $(0,0)$ and then compute the flux using the following snippet:

```
1  Fx = -y                              # force in the x-direction
2  Fy =  x                              # force in the y-direction
3  Fv = sp.Matrix([Fx, Fy])                      # force vector
4
5  x_ = R*sp.cos(θ)                 # Cartesian-polar mapping
6  y_ = R*sp.sin(θ)     # y0=0, semi-circle center is moved to (0,0)
7
8  dnds = sp.Matrix([y_.diff(θ), -x_.diff(θ)])
9
10 # Use the semi-circle curve with center at (0,0):
11 Q_C2 = sp.integrate((Gv.T@dnds).subs(dic), (θ, sp.pi/2, -sp.pi/2))[0]
12 display(Math(f" \\text{{Flux of the spin field across C$_2$ with (0,0)
13          as its center, via dot-product formula = }} {latex(Q_C2)}"))
```

Flux of the spin field across C_2 with $(0,0)$ as its center   via   dot-product formula $= 0$

This time, we got zero because the spin field is spinning along the semi-circle curve. Nothing is flowing across it.

These examples have led to the following understanding:

> When evaluating the flux across a curve, both the direction of the vector field and the direction of the curve are critical in determining its value in addition to, of course, the size of the curve.

3.7 Helmholtz decomposition

The Helmholtz decomposition is regarded as the fundamental theorem of vector calculus. It states that any vector field $\mathbf{F}$ that is sufficiently smooth in simply connected three-dimensional domain can be decomposed into a curl-free component and a divergence-free component. The curl-free field can be written as the gradient of a scalar potential ϕ, and the divergence-free field can be written as the curl of a vector potential $\mathbf{A}$. Therefore, the Helmholtz decomposition can be expressed as

$$\mathbf{F}(x, y, z) = -\underbrace{\nabla \phi(x, y, z)}_{\text{curl-free}} + \underbrace{\nabla \times \mathbf{A}(x, y, z)}_{\text{divergence-free}} \tag{3.55}$$

where ϕ is the scalar potential and $\mathbf{A}$ is the vector potential for a given $\mathbf{F}$. Using Eq. (2.28), it is easy to confirm that $\nabla \phi(x, y, z)$ will indeed be curl-free, and using Eq. (2.29), it is trivial to confirm that $\nabla \times \mathbf{A}(x, y, z)$ is divergence-free.

Equation (3.55) is a set of partial differential equations. Finding the analytic forms of ϕ and $\mathbf{A}$ can be a challenge; hence, we often do it numerically. The following presents a code and procedure for finding ϕ and $\mathbf{A}$ numerically for a given 2D vector field, modified based on a suggestion from ChatGPT. It uses a technique called fast Fourier transform (FFT), which is a widely used method for solving PDEs (among other applications) [6]. We have not yet discussed it and will simply use the FFT method provided by SymPy. The key idea is to transform the given vector field into the frequency domain, solve for the scalar potential in the frequency domain, and then obtain the divergence component through inverse transformation. Finally, the curl-free component is the difference between the original field and the divergence-free field. The detailed procedure is described in the following section.

3.7.1 *Procedure to find scalar and vector potentials*

1. **Grid and vector field setup**

   - The code defines a 2D mesh grid using `numpy.meshgrid` with `nx` by `ny` points. `X` and `Y` are the coordinate mesh grids corresponding to the `x` and `y` coordinates, respectively.
   - The given vector field $\mathbf{F}$ is represented as a single 3D numpy array of shape `(nx, ny, 2)`, where the first two dimensions correspond to spatial dimensions and the last dimension represents the vector components F_x and F_y.

2. **FFT**

   - The Fourier transform of the vector field $\mathbf{F}$ is computed using `scipy.fftpack.fftn`. This transforms the vector field into the frequency domain, where operations like differentiation and integration become algebraic (multiplicative).

- The frequency components are calculated using `scipy.fftpack.fftfreq`. This gives the wavenumbers k_x and k_y for the x- and y-directions, respectively.

3. Calculate scalar potential ϕ

- In the frequency domain, the divergence (which requires differentiations) of the vector field is simply the dot product of the wavenumber vector **k** and the Fourier-transformed vector field components.
- The scalar potential ϕ is obtained by dividing the divergence by the squared wavenumber magnitude $k^2 = k_x^2 + k_y^2$. This step yields the Fourier transform of ϕ.

4. Calculate curl-free component (gradient of ϕ)

- The gradient of ϕ in the frequency domain is obtained by multiplying ϕ by the wavenumber components k_x and k_y.
- The curl-free component (the gradient of ϕ) is obtained by transforming back to the spatial domain using the inverse Fourier transform `ifftn`.

5. Calculate divergence-free component (curl of vector potential)

- The divergence-free component is calculated by subtracting the curl-free component from the original vector field in the frequency domain.
- Finally, transform back to the spatial domain to obtain the divergence-free component using `ifftn`.

6. Visualization

- The original vector field, curl-free component, and divergence-free component are plotted using `plt.quiver`, which displays vector fields as arrows on a grid.

Note that when FFT is used, it assumes that the field has periodicity. Therefore, care is needed as violating this assumption can result in errors. It is always good practice to incorporate means in the procedure to check the solution obtained.

Let us take a look at some examples. First, a Python function is given as follows:

```python
1  from scipy.fftpack import fftn, ifftn, fftfreq   #needed for this method
2
3  def helmholtz_decomposition_2d(F, dx):
4      '''
5      Perform a numerical Helmholtz decomposition of a given 2D field F,
6      using FFT. Code and procedure are suggested by ChatGPT.
7          F:  the vector field, assumed having a periodicity, and the
8              periodic unit for F is discretized in the domain for FFT.
9          dx: Grid spacing in the domain.
10     '''
11     nx, ny, _ = F.shape                          # get the size of the grid
12     F_hat = fftn(F, axes=(0, 1))                   # Fourier Transform of F
13
14     kx = fftfreq(nx, d=dx)                       # Frequency domain coordinates
15     ky = fftfreq(ny, d=dx)
16     kx, ky = np.meshgrid(kx, ky, indexing='ij')
17     k2 = kx**2 + ky**2
18     k2[0, 0] = np.inf                            # To avoid division by zero
19
20     # Scalar potential phi, φ
21     div_F_hat = kx * F_hat[..., 0] + ky * F_hat[..., 1]
22     phi_hat = div_F_hat / k2
23
24     # Gradient of the scalar potential (curl-free component)
25     grad_φ_hat_x = kx * phi_hat
26     grad_φ_hat_y = ky * phi_hat
27     grad_φ = np.zeros_like(F)                        # inverse FFT
28     grad_φ[..., 0] = np.real(ifftn(grad_φ_hat_x, axes=(0, 1)))
29     grad_φ[..., 1] = np.real(ifftn(grad_φ_hat_y, axes=(0, 1)))
30
31     # Divergence-free component:
32     curl_A_hat=F_hat-np.stack([grad_φ_hat_x, grad_φ_hat_y], axis=-1)
33     curl_A = np.real(ifftn(curl_A_hat, axes=(0, 1)))
34
35     return grad_φ, curl_A
```

3.7.2 Example: Helmholtz decomposition of a vector field

This example computes the Helmholtz decomposition of a vector field that is a combination of a radial field and a spin field:

$$\mathbf{F}_{rs} = \begin{bmatrix} x \\ y \end{bmatrix} + \begin{bmatrix} -y \\ x \end{bmatrix} = \begin{bmatrix} x - y \\ y + x \end{bmatrix} \tag{3.56}$$

For this simple field, we know the expected results for both the curl-free and divergence-free fields. Therefore, it serves as a good test for our code. We will use `helmholtz_decomposition_2d()` to perform the decomposition. The following snippets define a standard square domain and then execute the decomposition:

```
1   # Examples: 2D vector field F (2D grid, 2 components)
2   nx, ny = 32, 32 # 64, 64    # mesh grid density, multiples of 8 for FFT
3   d = 1 # np.pi/2
4   x = np.linspace(-d, d, nx)                        # domain of F
5   y = np.linspace(-d, d, nx)
6   X, Y = np.meshgrid(x, y, indexing='ij')          # generate mesh grid
7
8   # The given vector field (you may use your own vector field)
9   F = np.zeros((nx, ny, 2))
10  F[..., 0] = X-Y # np.sin(Y)          # an r + s field, F_x component
11  F[..., 1] = X+Y # np.cos(X)                       # F_y component
12
13  # Decomposition using FFT:
14  dx = x[1] - x[0]; dy = y[1] - y[0]                # Grid spacing
15  grad_ϕ, curl_A = helmholtz_decomposition_2d(F, dx)
16  F_recov = grad_ϕ + curl_A            # compute the recovered field
```

The following Python function plots all four vector fields together: the original field, the curl-free component, the divergence-free component, and the recovered field, which is the sum of the curl-free and divergence-free components:

```
1    # Function to plot the vector fields:
2    def plot_vector_field(X, Y, U, V, k, title, scale=5):
3        plt.subplot(2, 2, k)
4        mag = np.sqrt(U**2 + V**2)               # to normalize the arrows
5        plt.quiver(X, Y, U, V, mag, scale=scale, scale_units='xy')
6        plt.title(title)                            #, cmap=cm.jet)
7        if k==3 and k==4: plt.xlabel('x')
8        if k==1 and k==3: plt.ylabel('y')
9        plt.axis('equal')
10       plt.grid(True)
```

The following snippet plots these four figures:

```
1  from matplotlib import cm                        # needed for the plots
2  plt.figure(figsize=(6, 6))                             # plot the results
3  plt.ioff() #plt.ion()#
4  plot_vector_field(X, Y, F[..., 0], F[..., 1], \
5                    1, 'Original Field', scale=5)
6  plot_vector_field(X, Y, F_recov[..., 0], F_recov[..., 1], \
7                    2, 'Recovered Field', scale=5)
8
9  plot_vector_field(X, Y, grad_ϕ[..., 0], grad_ϕ[..., 1], \
10                   3, 'Curl-Free', scale=5)
11 plot_vector_field(X, Y, curl_A[..., 0], curl_A[..., 1], \
12                   4, 'Divergence-Free', scale=5)
13
14 plt.savefig('imagesVC/HelmDecomp_rs.png', dpi=500)        # save to file
15 #plt.show()
```

These four vector fields, original, curl-free, divergence-free, and the recovered fields, are plotted in Fig. 3.13.

As shown clearly, we obtained the correct distribution for both the curl-free and divergence-free vector fields. The sum of these components recovers the original vector field.

To check the accuracy of the decomposed fields quantitatively, we compute the maximum curl of the curl-free field and the maximum divergence of the divergence-free field in the domain. We expect these values to be small and to decrease as the mesh grid density increases. The computation is carried out in the following steps:

1. Compute the norm of the original field in the domain using

$$\|\mathbf{F}\| = \sqrt{F_x^2 + F_y^2} \qquad (3.57)$$

This is used as the base for computing the relative error.

2. Compute the maximum curl and divergence at a point in the domain using

$$\text{curl}_{\max} = \max |\text{curl}(\mathbf{grad\_}\phi)|$$
$$\text{div}_{\max} = \max |\text{div}(\mathbf{curl\_A})| \qquad (3.58)$$

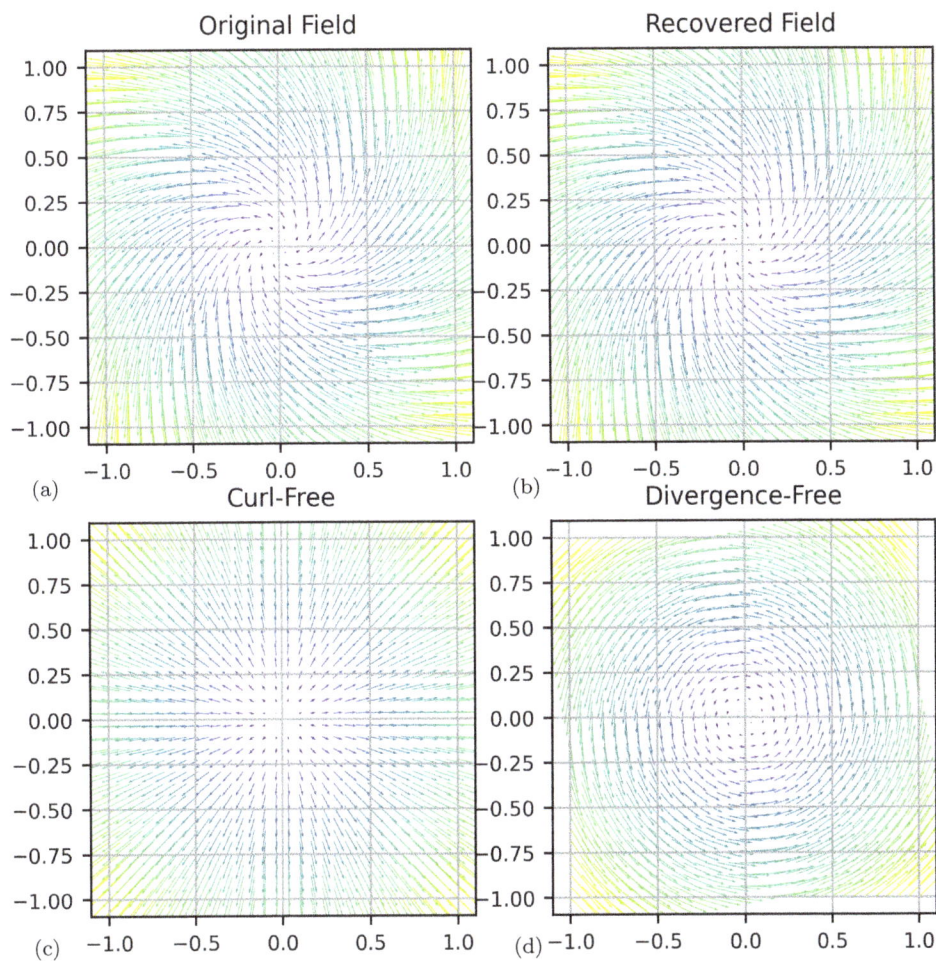

Figure 3.13. Results of the Helmholtz decomposition of a combined radial and spin field: (a) original, (b) recovered, (c) curl-free, and (d) divergence-free fields.

3. Compute the maximum relative error E using

$$E_{\text{curl-free}} = \frac{\text{curl}_{\text{max}}}{\text{mean}(\|\mathbf{F}\|)}$$

$$E_{\text{divergence-free}} = \frac{\text{div}_{\text{max}}}{\text{mean}(\|\mathbf{F}\|)}$$

$$(3.59)$$

The following are two utility functions for the above-mentioned computations:

```
 1  def compute_curl_2d(F, dx):
 2      # Compute the partial derivatives
 3      dFy_dx = np.gradient(F[..., 1], dx, axis=0)
 4      dFx_dy = np.gradient(F[..., 0], dx, axis=1)
 5      curl_F = dFy_dx - dFx_dy                          # Curl in 2D
 6
 7      return curl_F
 8
 9  def compute_divergence_2d(F, dx):
10      # Compute the partial derivatives
11      dFx_dx = np.gradient(F[..., 0], dx, axis=0)
12      dFy_dy = np.gradient(F[..., 1], dx, axis=1)
13      div_F = dFx_dx + dFy_dy                           # Divergence
14
15      return div_F
```

The snippet for computing the errors is given as follows:

```
 1  # Compute relative errors:
 2  norm_F = np.sqrt(F[..., 0]**2 + F[..., 1]**2)
 3  mean_norm_F = np.mean(norm_F)                  # the base for comparison
 4
 5  curl_of_curl_free = compute_curl_2d(grad_ɸ, dx)
 6  curl_max = np.max(np.abs(curl_of_curl_free))
 7
 8  div_of_div_free = compute_divergence_2d(curl_A, dx)
 9  div_max  = np.max(np.abs(div_of_div_free))
10
11  E_curl_free = curl_max/ mean_norm_F
12  E_div_free = div_max/ mean_norm_F
13
14  # Output the results
15  print("Relative Error for the curl-free component = ", E_curl_free)
16  print("Relative Error for the divergence-free component = ", E_div_free)
```

```
Relative Error for the curl-free component =  3.082170560448474e-15
Relative Error for the divergence-free component =  4.623255840672711e-15
```

The error level found is close to machine accuracy.

Additionally, we know that the curl-free and divergence-free components should be orthogonal. This can be verified using the following procedure:

$$\langle \mathbf{F}_{\text{curl-free}}, \mathbf{F}_{\text{div-free}} \rangle = \int \mathbf{F}_{\text{curl-free}} \cdot \mathbf{F}_{\text{div-free}} \, dV = 0 \qquad (3.60)$$

where $\mathbf{F}_{\text{curl-free}}$ is the curl-free component field and $\mathbf{F}_{\text{div-free}}$ is the divergence-free component field. This means that the dot product of the curl-free and divergence-free components, integrated over the entire domain, should equal zero.

The following snippet checks this:

```
1  # Compute the dot product at each grid point
2  dot_product = np.sum(grad_ϕ*curl_A, axis=-1)
3
4  # Integrate over the domain (sum all values, multiply by dx*dy)
5  orthogonality = np.sum(dot_product) * dx*dy
6
7  print(" Orthogonality check (close to zero?):", orthogonality)
```

```
Orthogonality check (close to zero?): 4.285272420182241e-16
```

As seen, the Helmholtz decomposition has successfully separated the mixed radial and spin fields, achieving accuracy close to machine precision.

Now, let's decompose a more complex field.

3.7.3 *Example: Helmholtz decomposition of a sinusoidal field*

In this example, we compute the Helmholtz decomposition of a sinusoidal field composed of sine and cosine functions:

$$\mathbf{F}_{sc} = \begin{bmatrix} \sin(y)\cos(x) \\ \cos(x)\sin(y) \end{bmatrix} \tag{3.61}$$

For this field, we do not have the exact forms for either the curl-free or divergence-free components. Let's use the following snippet to perform the Helmholtz decomposition numerically:

```
1  # Examples: 2D vector field F (2D grid, 2 components)
2  nx, ny = 256,256 #1024, 1024 # 512, 512 #128, 128 #32, 32 #64, 64 #
3  d = 2*np.pi
4  x = np.linspace(0, d, nx)              # domain of F, must match the
5  y = np.linspace(0, d, ny)              # periodicity of the field
6  X, Y = np.meshgrid(x, y, indexing='ij')        # generate mesh grid
7
8  # The given vector field (you may use your own vector field)
9  F = np.zeros((nx, ny, 2))
10 F[..., 0] = np.sin(Y) * np.cos(X) # np.sin(Y)
11 F[..., 1] = np.cos(X) * np.sin(Y) # np.cos(X)
12
13 # Numerical decomposition
14 dx = x[1] - x[0]                       # Grid spacing needed for FFT
15 grad_ϕ, curl_A = helmholtz_decomposition_2d(F, dx)
16 F_recov = grad_ϕ + curl_A
```

```
 1  plt.figure(figsize=(6, 6))                         # plot the results
 2  plt.ioff() #plt.ion()#
 3
 4  hx = int((nx+1)/2); hy = int((ny+1)/2)
 5  k = 4              # skip points, reduce the number of arrows to print
 6  Xh = X[:hx:k, :hy:k]; Yh = Y[:hx:k, :hy:k]
 7  plot_vector_field(Xh, Yh, F[:hx:k, :hy:k, 0], F[:hx:k, :hy:k, 1], \
 8                    1, 'Original Field',  scale=2)
 9  plot_vector_field(Xh, Yh,F_recov[:hx:k,:hy:k,0],F_recov[:hx:k,:hy:k,1],\
10                    2, 'Recovered Field', scale=2)
11
12  plot_vector_field(Xh, Yh,grad_ϕ[:hx:k,:hy:k,0],grad_ϕ[:hx:k,:hy:k,1],\
13                    3, 'Curl-Free',         scale=1)
14  plot_vector_field(Xh, Yh, curl_A[:hx:k,:hy:k,0],curl_A[:hx:k,:hy:k,1],\
15                    4, 'Divergence-Free', scale=1)
16
17  plt.savefig('imagesVC/HelmDecomp_sc.png', dpi=500)      # save to file
18  #plt.show()
```

These four vector fields, original, curl-free, divergence-free, and the recovered fields, are plotted in Fig. 3.14.

The following snippet computes the relative error of the decomposed fields:

```
 1  # Compute relative errors:
 2  norm_F = np.sqrt(F[..., 0]**2 + F[..., 1]**2)
 3  mean_norm_F = np.mean(norm_F)                # the base for comparison
 4
 5  curl_of_curl_free = compute_curl_2d(grad_ϕ, dx)
 6  curl_max = np.max(np.abs(curl_of_curl_free))
 7
 8  div_of_div_free = compute_divergence_2d(curl_A, dx)
 9  div_max  = np.max(np.abs(div_of_div_free))
10
11  E_curl_free = curl_max/ mean_norm_F
12  E_div_free = div_max/ mean_norm_F
13
14  # Output the results
15  print("Relative Error for the curl-free component = ", E_curl_free)
16  print("Relative Error for the divergence-free component = ", E_div_free)
```

```
Relative Error for the curl-free component =  0.010610733652569617
Relative Error for the divergence-free component =  0.010610733652571588
```

For this sinusoidal vector field, the error level is higher around 1.06% with a mesh grid of 256×256. This error can be reduced by increasing the

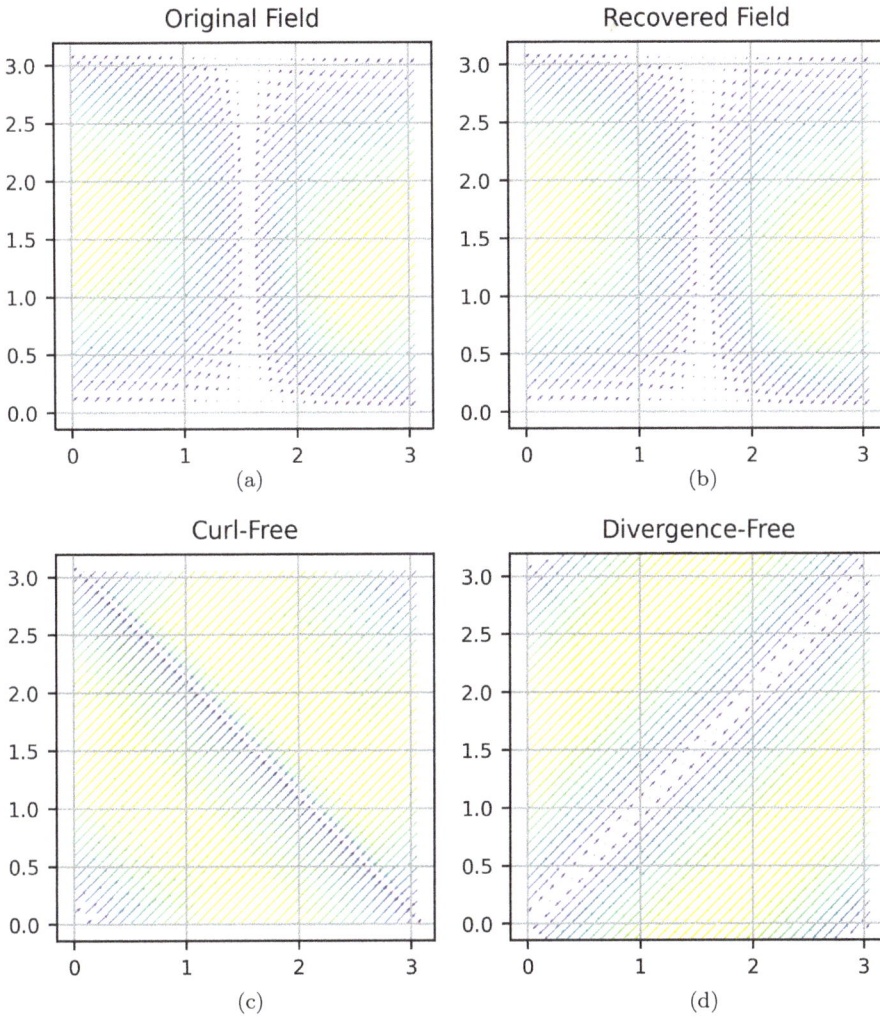

Figure 3.14. Results of the Helmholtz decomposition of sinusoidal field: (a) original, (b) recovered, (c) curl-free, and (d) divergence-free fields. The curl-free and divergence-free fields are orthogonal.

mesh density. When using a denser grid of 512×512, the error decreases to 0.53%. Further increasing the grid to 1024×1024 reduces the error to 0.267%. Readers are encouraged to experiment with different grid densities. Note that the computation is very fast — even with a grid of 1024×1024, it took less than a second to complete on the author's old laptop.

Next, let's verify the orthogonality of the curl-free and divergence-free components:

```
1  # Compute the dot product at each grid point
2  dot_product = np.sum(grad_φ*curl_A, axis=-1)
3
4  # Integrate over the domain (sum all values, multiply by dx*dy)
5  orthogonality = np.sum(dot_product) * dx*dy
6
7  print(" Orthogonality check (close to zero?):", orthogonality)
```

Orthogonality check (close to zero?): 1.4784748852176195e-11

As seen, the orthogonality is quite good and the Helmholtz decomposition has successfully separated the sinusoidal vector field.

The examples above demonstrate that the Helmholtz decomposition can be performed on smooth vector fields. Readers are encouraged to decompose other fields using the provided code. Since we employ a numerical method, the complexity of the vector field is not a significant issue as long as the functions are sufficiently smooth. However, it's essential to verify the numerical solutions obtained.

It's important to note that numerical studies using examples do not constitute a proof of the theorem. No matter how many cases are examined positively, we cannot conclusively prove that the theorem holds true in all cases. A rigorous theoretical proof for the Helmholtz decomposition and its uniqueness can be found at Helmholtz decomposition (https://en.wikipedia.org/wiki/Helmholtz_decomposition).

Intuitively, the Helmholtz decomposition theorem makes good sense. We know that any vector can be decomposed into two orthogonal vector components. There should be no reason why a vector function cannot be decomposed into two orthogonal vector component functions (curl-free and divergence-free), provided the functions are smooth enough for differentiation.

3.8 Concluding remarks

This chapter explores in detail the curve integrals of 2D vector functions. Various formulas, techniques, and Python codes have been presented for analyzing the effects of a vector function either along or across a curve within the vector function's domain. In particular, two major effects are examined: work and flux. The former measures the effects of a gradient field, which is curl-free, while the latter measures the effects of a solenoidal field, which is divergence-free. We also discussed the techniques for the Helmholtz decomposition of an arbitrary field into orthogonal curl-free and divergence-free components. The chapter concludes with the following remarks:

1. The work done by a vector field along a curve is obtained by integrating the dot product of the vector field with the tangent vector along the curve. This results in a scalar integrand that can be evaluated using parameterization techniques.
2. The flux of a vector field across a curve is obtained by integrating the dot product of the vector field with the normal vector on the curve. This also results in a scalar integrand that can be evaluated using parameterization techniques.
3. The work and flux measure mathematically two important features of a given vector field relative to a given curve, linking directly to mass conservation laws.
4. Both the direction of the vector field and the direction of the curve are critical in determining the effects alongside the size of the curve.
5. There are special and useful fields, such as gradient fields, which have potential functions where the gradient of the potential function is the vector field itself. A gradient field is conservative, implying the work done by the field on a moving particle depends only on the initial and final points of the particle and is independent of the path taken. Any work done along a closed curve is zero.
6. In a conservative force field, the total energy (kinetic plus potential) of a moving particle remains constant.
7. A Python function has been provided to find the potential function of a given vector field, provided it is conservative.
8. Codes for the numerical Helmholtz decomposition of an arbitrary smooth field into orthogonal curl-free and divergence-free fields are provided along with methods for error assessment.

The following chapter will discuss how a curve integral is related to a domain integral for vector fields, specifically focusing on Green's theorems and conservation laws.

References

[1]   G. Strang, *Calculus*, Wellesley-Cambridge Press, Massachusetts, 1991.
[2]   G.R. Liu, *Numbers and Functions: Theory, Formulation, and Python Codes*, World Scientific, New Jersey, 2024.

[3]   G.R. Liu, *Mechanics of Materials: Formulations and Solutions with Python*, World Scientific, New Jersey, 2024.

[4]   G.R. Liu, *Calculus: A Practical Course with Python*, World Scientific, New Jersey, 2025.

[5]   G.R. Liu, *Machine Learning with Python: Theory and Applications*, World Scientific, New Jersey, 2023.

[6]   G.R. Liu and Z.C. Xi, *Elastic Waves in Anisotropic Laminates*, CRC Press, New York, 2001.

Chapter 4

Green's Theorems and Applications

```
1   # Place cursor in this cell, and press Ctrl+Enter to import dependences.
2   import sys                        # For accessing the computer system
3   sys.path.append('../grbin/')         # Add in the path to your system
4
5   from commonImports import *      # Import dependences from '../grbin/'
6   import grcodes as gr               # Import the module of the author
7   importlib.reload(gr)            # When grcodes is modified, reload it
8
9   init_printing(use_unicode=True)      # For latex-like quality printing
10  np.set_printoptions(precision=4,suppress=True,
11                      formatter={'float_kind': '{:.4e}'.format})
```

In the previous chapter, we explored the curve integrals of 2D vector functions in detail. We derived formulas, techniques, and Python codes to evaluate two major effects, work and flux, using curve integrals. This chapter discusses how a curve integral relates to a domain integral bounded by the curve for vector fields, focusing on Green's theorem and conservation laws.

Green's theorem has two versions: One relates the work done along a curve to the domain integral of the circulation of the vector field and the other relates the flux across a curve to the domain integral of the divergence of the vector field. The former corresponds to the energy conservation law, while the latter corresponds to the mass conservation law. For convenience in discussion, the former is referred to as Green's circulation theorem, and the latter is referred to as Green's divergence theorem. Both versions will be presented in this chapter, and their commonalities and differences will be examined.

Additionally, a technique and code for computing the area of a polygon with an arbitrary number of edges will be developed using Green's theorems.

We will also examine features of vector fields that are divergence-free, providing codes for finding the stream function, which complements the potential function discussed in the previous chapter. This will lead to a discussion on a very special type of field that is both conservative and divergence-free.

This chapter is written with reference to textbooks in Refs. [1–3]. Wikipedia pages, particularly those on integral (https://en.wikipedia.org/wiki/Integral), surface integral (https://en.wikipedia.org/wiki/Surface_integral), and Green's theorem (https://en.wikipedia.org/wiki/Green%27s_theorem), serve as valuable additional references.

Both NumPy and SymPy are used in the development of code for the demonstration examples. Discussions with ChatGPT, Gemini, and Bing have also greatly helped in coding and in the preparation of this chapter.

4.1 Green's theorem along boundary and over domain

We discussed the fundamental theorem of calculus in Chapter 4 of Ref. [4] when studying 1D integrals, where a 1D integral is converted into the values of the antiderivative at the two boundary points. Green's theorem extends this fundamental concept to 2D integrals by converting a 2D domain integral into a closed curve integral along its boundary.

Consider two arbitrary differentiable functions, $M(x, y)$ and $N(x, y)$, defined in a **simply connected domain** $\mathcal{D}$, which is bounded by a closed boundary curve $\mathcal{C}$. Green's theorem is expressed as

$$\oint_{\mathcal{C}} (M \, dx + N \, dy) = \iint_{\mathcal{D}} \left(\frac{\partial N}{\partial x} - \frac{\partial M}{\partial y} \right) dx \, dy \qquad (4.1)$$

In general, we also require all the integrands involved to be integrable. Since the functions $M(x, y)$ and $N(x, y)$ are assumed to be differentiable, these integrands are typically (Riemann) integrable although we may not always be able to find the analytical forms of their antiderivatives.

Since the domain integral on the right side of Eq. (4.1) represents the circulation, Eq. (4.1) is referred to as **Green's circulation theorem**.

4.1.1 *Conditions on the domain and its boundary curve*

Green's theorem holds under the following conditions:

1. **Simply connected domains:** The theorem applies to simply connected domains. A domain is said to be simply connected if any closed curve within it can shrink to a point without leaving the domain. Essentially, this means that the domain has no holes and can be either convex or concave.
2. **Piecewise smooth boundary curve:** The domain must have a piecewise smooth boundary curve. This requires that the boundary is formed by a finite number of smooth segments without gaps or overlaps. Each segment is smooth though the smoothness at the connecting points is not required; however, the curve must remain continuous.
3. **Extension to 3D:** The concept of piecewise smooth boundaries naturally extends to 3D problems, where the domain has a piecewise smooth boundary surface. This surface consists of a finite number of smooth surface segments, and while each segment is smooth, the connecting curves do not need to be smooth, but the surface must be continuous. This condition will also be used in Chapter 6 on 3D problems.

Additional notes:

- The functions $M(x, y)$ and $N(x, y)$ are arbitrary and need not be related to a vector function. The only requirement is that they are (piecewise) differentiable. While they can be component functions of a vector field, this is not a necessity.

4.1.2 *Proof of Green's theorem*

The proof is done in two steps. First, we proof

$$\oint_C M \, dx = \iint_D \left(-\frac{\partial M}{\partial y} \right) dx \, dy \tag{4.2}$$

and then use the similar trick and procedure to proof

$$\oint_C N \, dy = \iint_D \frac{\partial N}{\partial x} dx \, dy \tag{4.3}$$

Adding up the two proven expressions for $M(x, y)$ and $N(x, y)$ leads to Green's theorem, as shown in Eq. (4.1). This supports our argument that $M(x, y)$ and $N(x, y)$ do not necessarily need to be related, and Eqs. (4.2) and (4.3) hold individually.

Intuitive understanding of the equations: There is a clear pattern in these two equations that aids in the intuitive understanding. The double integral in Eq. (4.2) for $M(x, y)$ over the 2D domain is first integrated over y, leaving a single integral over x. Since this integral is now along the boundary curve of the 2D domain, a similar process is observed in Eq. (4.3), where the double integral of $N(x, y)$ is first integrated over x, leaving a single integral over y along the boundary curve.

Rigorous proof: For a more rigorous proof, consider the case where the domain is convex. The boundary curve that bounds the bottom part of the domain $\mathcal{D}$ is denoted as $\mathcal{C}_B$, with the curve function $S_B(x)$. The boundary curve that bounds the top part of $\mathcal{D}$ is $\mathcal{C}_T$, with the curve function $S_T(x)$, as illustrated in Fig. 4.1(a).

The double integral on the right side of Eq. (4.2) becomes

$$\iint_{\mathcal{D}} \left(-\frac{\partial M}{\partial y} \right) dx\, dy = \int_a^b - \left[\int_{S_B(x)}^{S_T(x)} \frac{\partial M}{\partial y}(x, y)\, dy \right] dx$$

$$= \int_a^b \left[-M(x, S_T(x)) + M(x, S_B(x)) \right] dx \qquad (4.4)$$

$$= \int_a^b M(x, S_B(x))\, dx + \int_b^a M(x, S_T(x))\, dx$$

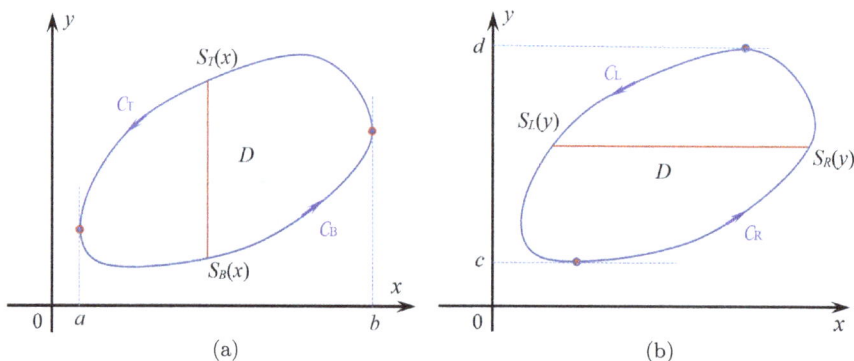

Figure 4.1. Integral procedures for the proof of Green's theorem: (a) integrate along y first and then finish up along x and (b) integrate along x first and then finish up along y.

Standard sign convention: Note that we always follow the **standard sign convention** for line integration: The positive direction is counterclockwise, with the domain on the left-hand side.

Splitting the closed curve integral: The closed curve integral on the left side of Eq. (4.2) can be split into two separate curve integrals: one along the boundary curve C_B and one along the boundary curve C_T. This decomposition is illustrated as follows:

$$\oint_C M \, dx = \int_{C_B} M \, dx + \int_{C_T} M \, dx$$
$$= \int_a^b M(x, S_B(x)) \, dx + \int_b^a M(x, S_T(x)) \, dx \tag{4.5}$$

This is the same as that obtained in Eq. (4.4). Hence, Eq. (4.2) is proven.

To prove Eq. (4.3), we use a similar technique. We now consider the left boundary of $\mathcal{D}$ as C_L with boundary function $S_L(y)$, and the right boundary of $\mathcal{D}$ as C_R with boundary function $S_R(y)$, as depicted in Fig. 4.1(b). The double integral on the right side of Eq. (4.3) can be transformed into

$$\iint_{\mathcal{D}} \frac{\partial N}{\partial x} \, dx \, dy = \int_c^d \left[\int_{S_L(y)}^{S_R(y)} \frac{\partial N}{\partial x}(x, y) \, dx \right] dy$$
$$= \int_c^d \left[N(S_R(y), y) - N(S_L(y), y) \right] dy \tag{4.6}$$
$$= \int_c^d \left[N(S_R(y), y) + \int_d^c N(S_L(y), y) \right] dy$$

Next, the curve integral on the left side of Eq. (4.3) is

$$\oint_C N \, dy = \int_{C_R} N \, dy + \int_{C_L} N \, dy$$
$$= \int_c^d N(S_R(y), y) \, dy + \int_d^c N(S_L(y), y) \, dy \tag{4.7}$$

This is the same result obtained in Eq. (4.6), thus proving Eq. (4.3). This completes the proof of Green's theorem under the assumption that the domain is convex.

4.1.3 *Sub-domain divisions*

The above proof assumes that domain $\mathcal{D}$ is convex, which allows us to perform integrals along a single direction without leaving the domain. For concave domains, however, we need to divide the domain into convex sub-domains, as illustrated in Fig. 4.2(a).

Concave to convex: The splitting process always allows Green's theorem to be applied to each convex sub-domain. The theorem's additive nature and consistent sign conventions ensure that integrating over the entire concave domain yields the same result as integrating over the individual convex sub-domains. When integrating along cutting lines, contributions in opposite directions cancel out, as illustrated in Fig. 4.2(a). This splitting is primarily a conceptual tool to prove the validity of Green's theorem for concave domains. In practical computations, such a division may not always be necessary.

Formation of simply connected sub-domains: For domains with holes, which do not satisfy the conditions for Green's theorem, the domain must be divided into simply connected sub-domains, as shown in Fig. 4.2(b). Green's theorem can then be applied individually to each simply connected sub-domain. Integrals along boundaries of holes are computed separately and canceled out appropriately, following the standard sign convention: positive in the counterclockwise direction with the domain on the left-hand side.

Domain sub-division for piecewise-differentiable functions: If the function M or N is piecewise differentiable, the integral domain should be sub-divided along the boundary separating the pieces so that within each sub-domain, the function is differentiable. In this case, we still require

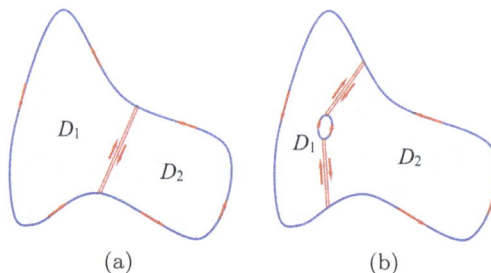

(a)   (b)

Figure 4.2. Sub-domain divisions: (a) a concave domain split into two convex sub-domains and (b) a domain with a hole divided into two simply connected sub-domains.

the function to be continuous across the entire domain. This domain sub-division technique applies to all piecewise-differentiable functions involved in integrals, including line, curve, surface, and domain integrals.

The line integral on the left side of Eq. (4.1) can be evaluated using parameterization techniques, as discussed in Chapter 3. This integral represents the work done by the vector field.

The domain integral on the right side can be computed using techniques from Chapter 5 of Ref. [4]. Since Green's theorem equates these two integrals, the simpler method can be selected based on the specific problem at hand.

4.1.4 *Physical meaning of Green's theorem*

The physical meaning of these integrals depends on the type of problem. For example, if the vector field $\mathbf{V}$ is a force field $\mathbf{F}$ with two component functions F_x and F_y, Green's theorem can be expressed as

$$\oint_C (F_x \, dx + F_y \, dy) = \iint_D \left(\frac{\partial F_y}{\partial x} - \frac{\partial F_x}{\partial y} \right) dx \, dy \qquad (4.8)$$

As discussed in Chapter 3, the left side of Eq. (4.8) is the work done by the force field along the closed curve C that bounds the domain D. The question is what the term on the right side of Eq. (4.8) represents.

4.1.5 *Curl of a force field*

As discussed in Section 2.8, the right side of Eq. (4.8) represents the third component of curl $\mathbf{F}$, which corresponds to the circulation of the force field with respect to the z-axis. Since we are considering the 2D (x, y) plane, the only relevant curl component for Eq. (4.8) is the curl in the (x, y) plane (with respect to the z-axis). The physical meaning of Green's theorem for a force field can be summarized as

$$\underbrace{\oint_C (F_x \, dx + F_y \, dy)}_{\text{total work along } C} = \underbrace{\iint_D \left(\frac{\partial F_y}{\partial x} - \frac{\partial F_x}{\partial y} \right) dx \, dy}_{\text{total circulation in } D} \qquad (4.9)$$

A simple example is given in Chapter 2, where the curl of a velocity vector in the 2D (x, y) plane is twice the angular velocity with respect to the z-axis. Another example can be found in Chapter 9 of Ref. [3] for a long bar with a circular cross-section subjected to a torque, where the curl of the shear stress field is twice the twist angle of the bar (for unit shear modulus).

4.1.6 *A pictorial proof of Green's circulation theorem*

We proved Green's circulation theorem, Eq. (4.1). This section provides a pictorial description to explicitly show the insight into the cancellation of the circulation of a vector function within the integral domain, leaving only an integral of the vector function along the boundary curve.

Figure 4.3 shows a discretized domain $\mathcal{D}$ bounded by $\mathcal{C}$ with small cells. The arrows represent the circulations along each of the cell edges. It is clear that all red and blue arrows on the same edge will cancel each other out because each pair of red and blue arrows has the same value (although we do not know the exact value) but opposite directions. This cancellation occurs along all internal edges, leaving only the green arrows along the edges on $\mathcal{C}$, as shown in Fig. 4.3.

Based on Fig. 4.3, the rigorous proof given in Section 4.1 can easily be applied to each of the sub-domains formed by a column of cells (or a row of cells). Note that if the vector function is piecewise differentiable, the line that divides two of the pieces should be one of the sub-division lines in Fig. 4.3.

4.1.7 *Code for examining Green's circulation theorem*

To efficiently examine Green's circulation theorem for different vector fields, we write the following Python function. The tasks this function aims to perform are listed in its docstring:

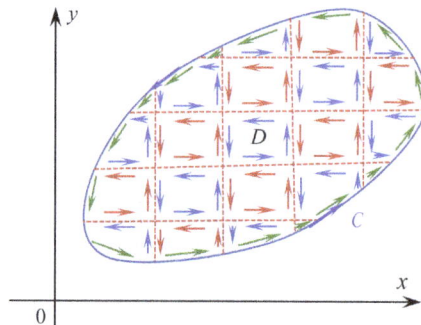

Figure 4.3. Cancellation of the circulation of a vector function within the integral domain, leaving an integral of the vector function along the boundary curve.

```
 1  def exam_GreensTheorem1(V, X, a):
 2      '''Examine a given vector field V that is differentiable over
 3          a circular domain. The code prints out:
 4          1 The circulation of the vector field V.
 5          2 The total circulation of the field over the circle.
 6          3 The total work by V along the boundary curve of the circle.
 7          4 Whether or not the Green's circulation theorem holds.
 8          The size of the circular domain is controlled by the radius a.
 9      '''
10      r, θ = symbols('r, θ', real=True) # for integration in polar coord.
11      x, y = X                          # unpack the Cartesian coordinates
12
13      # Coordinate transformation, for integral on the closed circle C:
14      xC = a*cos(θ);   yC = a*sin(θ)          # xC, yC are x,y on curve C
15      dicC = {x:xC, y:yC}
16
17      tds = Matrix([xC.diff(θ), yC.diff(θ)])              # tds = dr
18      fIC = V.T@tds                       # integrand for the curve integral
19      fIC = fIC.subs(dicC)
20      display(Math(f"\\text{{ Integrand of the curve integral}}=\
21                          {latex(sp.simplify(fIC)[0])}"))
22      Int = integrate
23      Vts = Int(fIC, (θ,0,2*sp.pi)).simplify()[0]
24
25      # Integral over the circular domain D:
26      xD = r*cos(θ)                   # xD, yD are x,y in the circular Domain
27      yD = r*sin(θ)
28      dicD = {x:xD, y:yD}
29      curl_V = gr.curl_vf(V, X).simplify()      # the curl in the Domain
30      display(Math(f" \\text{{ curl of the field}}={latex(curl_V)}"))
31
32      fID = (curl_V*r).simplify()  # integrand in circular domain, detJ=r
33      circulVD = Int(Int(fID.subs(dicD),(θ,0,2*sp.pi)),(r,0,a)).simplify()
34
35      # print out the results:
36      display(Math(f" \\text{{Total work of the vector field along the \
37                          circular boundary}} = {latex(Vts)}"))
38      display(Math(f" \\text{{Total circulation over the circular domain\
39                          }}= {latex(circulVD)}"))
40      print(f" Does Green's circulation theorem hold? {Vts==circulVD}")
```

4.1.8 *Example: Position fields with zero circulation*

Consider the following position fields:

$$\mathbf{r}_1 = \begin{bmatrix} x \\ y \end{bmatrix}; \quad \mathbf{r}_2 = \begin{bmatrix} x \\ y + 2a \end{bmatrix} \tag{4.10}$$

where a is a given radius of the circle used to evaluate Green's theorem.

Show the following:

1. These fields are curl-free and have zero circulation.
2. The curve integral on the circular boundary centered at the origin equals the integral over the circular domain, and Green's theorem holds.

Solution: We use `exam_GreensTheorem1()` and the following code snippet to complete the tasks:

```
1  x, y = symbols('x, y ', real=True)              # define variables
2  a = symbols('a', nonnegative=True)              # define variables
3  X = sp.Matrix([x, y])                        # the 2D cooridnate vector
4
5  r1 = Matrix([x, y ])                           # the given vector field
6  exam_GreensTheorem1(r1, X, a)
```

Integrand of the curve integral $= 0$

Curl of the field $= 0$

Total work of the vector field along the circular boundary $= 0$

Total circulation over the circular domain $= 0$

Does Green's circulation theorem hold? True

```
1  r2 = Matrix([x, y+2*a ])                        # the given vector field
2  exam_GreensTheorem1(r2, X, a)
```

Integrand of the curve integral $= 2a^2 \cos(\theta)$

Curl of the field $= 0$

Total work of the vector field along the circular boundary $= 0$

Total circulation over the circular domain $= 0$

Does Green's circulation theorem hold? True

We found the following:

1. The circulation of the 2D position fields (radial flow) is zero because the curl in the entire space is zero.
2. The work done along the closed circle is also zero, which equals the total circulation within the domain. Green's theorem holds as $0 = 0$.

These findings also hold for r_2, implying that the center of the radial flow is immaterial in these cases. This is because these position fields yield zero curl throughout the entire space.

4.1.9 *Example: Examination of the 2D gravity field*

Consider the gravity field studied in Section 3.6.2. Here, we examine a circular domain and its boundary curve. Let us conduct the same examination as in the previous examples using the same code:

```
1  m, g = symbols('m, g', positive=True)        # mass and gravity
2
3  GFv = sp.Matrix([0*g, -m*g                    # vertical gravity field
4  exam_GreensTheorem1(GFv, X, a)
```

Integrand of the curve integral $= -agm\cos(\theta)$

Curl of the field $= 0$

Total work of the vector field along the circular boundary $= 0$

Total circulation over the circular domain $= 0$

Does Green's circulation theorem hold? True

We found zero curl and, hence, zero circulation because the gravitational field is constant, resulting in zero circulation throughout the 2D space. This leads to a zero integrand for the domain integration term. The curve integral is also zero (even though the integrand itself is nonzero) because the curve is enclosed and the gradient field is conservative, as discussed in Chapter 3. In any case, Green's theorem holds.

4.1.10 *Example: Green's theorem on spin fields*

Consider the spin fields

$$\mathbf{s}_z = \begin{bmatrix} -y \\ x \end{bmatrix}; \quad \mathbf{s}_{r1} = \begin{bmatrix} -\frac{y}{r} \\ \frac{x}{r} \end{bmatrix}; \quad \mathbf{s}_{r2} = \begin{bmatrix} -\frac{y}{r^2} \\ \frac{x}{r^2} \end{bmatrix} \tag{4.11}$$

where $r = \sqrt{x^2 + y^2}$.

For each of these fields, perform the following **tasks**:

1. Compute the curl of the field.
2. Compute the work done by the field along the circular boundary of a circular domain centered at the origin.
3. Compute the domain integral of the curl over the circular domain.
4. Check whether Green's circulation theorem holds.

Solution for the field $\mathbf{s}_z$: We use exam_GreensTheorem1() and the code snippet below to complete the tasks.

```
1  x, y = symbols('x, y', real=True)
2  a = symbols('a', positive=True)              # radius of the circle
3  X = Matrix([x, y])
4
5  sz = Matrix([-y, x])                          # the given vector field
6  exam_GreensTheorem1(sz, X, a)
```

Integrand of the curve integral $= a^2$

Curl of the field $= 2$

Total work of the vector field along the circular boundary $= 2\pi a^2$

Total circulation over the circular domain $= 2\pi a^2$

Does Green's circulation theorem hold? True

As seen, the spin field has a constant curl of 2. The work done by the spin field over any closed circular boundary is twice the area of the circle, which corresponds to the total circulation within the circle, confirming that Green's theorem holds. The spin flow circulates along the circular boundary.

Solution for the field s_{r1}: This is the spin field s_z divided by r. We use exam_GreensTheorem1() and the following code snippet to complete the tasks:

```
1  s1x = -y/sp.sqrt(x**2+y**2)
2  s1y =  x/sp.sqrt(x**2+y**2)
3  V = Matrix([s1x, s1y])
4  exam_GreensTheorem1(V, X, a)
```

Integrand of the curve integral $= a$

Curl of the field $= \dfrac{1}{\sqrt{x^2 + y^2}}$

Total work of the vector field along the circular boundary $= 2\pi a$

Total circulation over the circular domain $= 2\pi a$

Does Green's circulation theorem hold? True

As seen, the curl of the spin field is $1/r$. The work done by the spin field over any closed circular boundary is twice the area of the circle divided by a, which corresponds to the total circulation within the circle, confirming that Green's theorem holds.

Solution for the field s_{r2}: This is the spin field s_z, which is not a gradient field, divided by r^2. However, we found in Example 3.4.8 that s_{r2} is a gradient field, and the potential function was determined to be $\phi(x, y) = \arctan\left(\frac{y}{x}\right)$. We use exam_GreensTheorem1() and the following code snippet to complete the tasks:

```
1  s2x = -y/(x**2+y**2)
2  s2y =  x/(x**2+y**2)
3  V = Matrix([s2x, s2y])
4  exam_GreensTheorem1(V, X, a)
```

Integrand of the curve integral $= 1$

Curl of the field $= 0$

Total work of the vector field along the circular boundary $= 2\pi$

Total circulation over the circular domain $= 0$

Does Green's circulation theorem hold? False

4.1.11 *Consequence of violating the differentiability condition*

This time, we found that Green's circulation theorem does not hold. The reason is that the field $s_{r2} = \left[-\frac{y}{r^2} \quad \frac{x}{r^2}\right]^T$ has a singularity (also called a pole) at $(0, 0)$. Due to this singularity, the vector field is not differentiable at the origin, violating the conditions we set for the theorem at the beginning. In the exam_GreensTheorem1() function, we computed its curl anyway, which incorrectly resulted in a zero value. This result is incorrect, as the theorem is not supposed to hold for nondifferentiable vector functions, and the code reflects this consequence.

Notably, the effect of the pole is correctly captured by the curve integral, which produces a value of 2π because the circular curve path used encloses the pole. If the domain $\mathcal{D}$ is placed anywhere in the domain excluding the origin, the vector function becomes differentiable, resulting in a zero curl and a zero domain integral. In such a case, the curve integral will also be zero, as the field is a gradient field, and the flux across any closed curve (that does not enclose any poles) will be zero, ensuring that Green's circulation theorem holds.

Readers are encouraged to modify the exam_GreensTheorem1() function to perform examinations using a circular domain that excludes the origin to further explore the behavior of the theorem under different conditions.

4.1.12 *A remark on Green's circulation theorem*

Finally, one may ask why the field $\mathbf{s}_{r1} = \left[-\frac{y}{r} \ \frac{x}{r}\right]^{\top}$ obeys Green's circulation theorem, even though it also has a pole at $r = 0$. The reason is that this pole is **removable**. In this case, the numerators x and y approach zero at the same rate as r approaches zero, effectively canceling out the singularity at the origin. This makes the field well behaved and differentiable everywhere except potentially at the origin, where the singularity is mitigated, allowing Green's circulation theorem to hold in this context [4].

4.1.13 *Example: Green's theorem for a mixed field*

Consider a field that is a mixture of the position and spin fields:

$$\mathbf{V} = \begin{bmatrix} x - y \\ y + x \end{bmatrix} \tag{4.12}$$

The snippet to examine it is as follows:

```
1  Vx, Vy = symbols('V_x, V_y', real=True)
2  Vx = x-y; Vy = y+x
3  V = Matrix([Vx, Vy])                          # an arbitrary field
4  exam_GreensTheorem1(V, X, a)
```

Integrand of the curve integral $= a^2$

Curl of the field $= 2$

Total work of the vector field along the circular boundary $= 2\pi a^2$

Total circulation over the circular domain $= 2\pi a^2$

Does Green's circulation theorem hold? True

It is seen that these integrals capture only the circulation part of the field.

4.1.14 *Example: Green's theorem on an arbitrary field*

Consider an arbitrarily constructed vector field:

$$\mathbf{V} = \begin{bmatrix} -x^2 y + x \\ x^2 y \end{bmatrix} \tag{4.13}$$

The code to examine it is as follows:

```
1  Vx, Vy = symbols('V_x, V_y', real=True)
2  Vx = -x**2*y+x; Vy = x**2
3  V = Matrix([Vx, Vy])                    # an arbitrary field
4  exam_GreensTheorem1(V, X, a)
```

Integrand of the curve integral $=$

$$a^2 \left(a \cos^2 (\theta) + \left(\frac{a^2 \sin (2\theta)}{2} - 1 \right) \sin (\theta) \right) \cos (\theta)$$

Curl of the field $= x (x + 2)$

Total work of the vector field along the circular boundary $= \dfrac{\pi a^4}{4}$

Total circulation over the circular domain $= \dfrac{\pi a^4}{4}$

`Does Green's circulation theorem hold? True`

The nonlinear vector field has a complex circulation that varies in the domain, but Green's theorem still holds. The curve integral captures the work done by the field along the circular boundary, while the domain integral captures the total circulation of the field. These two integrals are equal.

The examples presented above conclude that Green's circulation theorem holds for any differentiable vector functions.

4.2 Green's theorem application: Area computation

As mentioned earlier, the functions $N(x, y)$ and $M(x, y)$ are arbitrary, and the only condition is that they are differentiable. Hence, we can set them in desirable forms. Let $M(x, y) = -y$ and $N(x, y) = x$, which represents a spin field with a constant curl of 2. In this case, Green's theorem, Eq. (4.1), becomes

$$\oint_C (x \, dy - y \, dx) = 2 \underbrace{\iint_D dx \, dy}_{A} = 2A \tag{4.14}$$

Clearly, the right side is simply twice the area A of the domain $\mathcal{D}$. This means that the curl of the spin field is twice the area covered by the field, which is also the total circulation. This provides a good understanding of the curl of a vector field.

Now, the curve integral on the left side of Eq. (4.14) can be used to compute the area of a given domain.

Alternatively, one can also set $M(x, y) = 0$ and $N(x, y) = x$ or $M(x, y) = -y$ and $N(x, y) = 0$ to compute A. The formulas are, respectively,

$$A = \oint_C x \, dy$$

$$A = \oint_C -y \, dx$$

(4.15)

Let us look at some examples.

4.2.1 *Example: Area of a circle and an ellipse*

The following code computes the area of a circle based on Green's theorem using Eq. (4.15):

```
1  R = symbols('R ', positive=True)
2  x, θ = symbols('x, θ ', real=True)
3
4  # parameterize the circular curve with θ:
5  x = R*sp.cos(θ)                          # Cartesian-polar transformation
6  y = R*sp.sin(θ)
7
8  fI =-y*(-R*sin(θ))              # use A = ∮-ydx;    dx = (-R*sin(θ))dθ
9  #fI = x*(R*cos(θ))             # or use A = ∮ xdy;    dy = (R*cos(θ))dθ
10
11 A = sp.integrate(fI, (θ, 0, 2*sp.pi))
12 display(Math(f" \\text{{Area of the circle of radius R = }}{latex(A)}"))
```

Area of the circle of radius $R = \pi R^2$

The curve of the ellipse can be expressed using an angle parameter θ:

$$x = a\cos(\theta); \quad y = b\sin(\theta)$$

(4.16)

where a and b are the semi-major and semi-minor axes of the ellipse.

The code for computing the area of an ellipse is as simple as follows:

```
1  a, b = symbols('a, b', positive=True)
2  x, θ = symbols('x, θ ', real=True)
3
4  # parameterize the ellipic curve with θ:
5  x = a*sp.cos(θ)                          # dx = (-a*sin(θ))dθ
6  y = b*sp.sin(θ)                          # dy = (b*cos(θ))dθ
7
8  fI =(x*(b*cos(θ)) - y*(-a*sin(θ)))/2
9  A = sp.integrate(fI, (θ, 0, 2*sp.pi))
10 display(Math(f" \\text{{Area of the ellipse with radia a and b = }}\
11                                        {latex(A)}"))
```

Area of the ellipse with radia a and $b = \pi ab$

4.2.2 *Formula for the area of a polygon*

Since a polygon is composed of straight edges connected together, the integrals in Eq. (4.14) can be converted to a discrete form using the vertices or nodes of the polygon. The process is as follows.

Consider a polygon with n straight edges numbered counterclockwise, with the polygon oriented as shown. The numbering of the nodes and edges follows the same convention, as illustrated in Fig. 4.4.

Equation (4.14) can then be written as

$$A = \frac{1}{2} \oint_C (x \, dy - y \, dx)$$

$$= \frac{1}{2} \sum_{i=1}^{n} \left(\underbrace{\frac{x_{i+1} + x_i}{2}}_{\text{average } x \text{ on edge } i} (dy)_i - \underbrace{\frac{y_{i+1} + y_i}{2}}_{\text{average } y \text{ on edge } i} (dx)_i \right) \qquad (4.17)$$

where n is the number of edges of the polygon. For the ith edge, because it is straight, we have

$$(dy)_i = y_{i+1} - y_i; \quad (dx)_i = x_{i+1} - x_i \qquad (4.18)$$

This relation is clearly shown in Fig. 4.4.

Substituting Eq. (4.18) into Eq. (4.17), we have

$$A = \frac{1}{2} \sum_{i=1}^{n} \left(\frac{x_{i+1} + x_i}{2} (y_{i+1} - y_i) - \frac{y_{i+1} + y_i}{2} (x_{i+1} - x_i) \right)$$

$$= \frac{1}{2} \sum_{i=1}^{n} (x_i y_{i+1} - y_i x_{i+1}) \qquad (4.19)$$

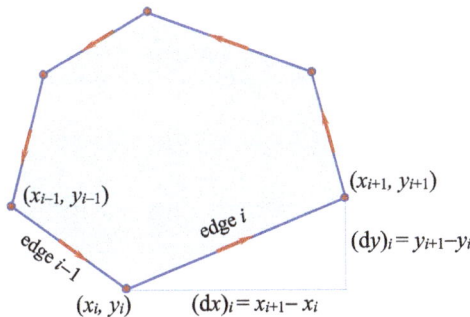

Figure 4.4. A polygon with all straight edges. Its nodes are numbered counterclockwise. The differences dy and dx for an edge can be evaluated using the differences in coordinates.

The derivation leading to the final result in Eq. (4.19) is found using the following snippet:

```
1  xi, xi1, yi, yi1 = symbols('x_i, x_i1, y_i, y_i1', real=True)
2  ((xi + xi1)*(yi1 - yi)/2 - (yi + yi1)*(xi1 - xi)/2).simplify()
```

$$x_i y_{i1} - x_{i1} y_i$$

A Python function using Eq. (4.19) is given in the following section.

4.2.3 *Code for the area of arbitrary polygons*

```
1  def polygon_area_green(vertices):
2      """
3      Compute the area of a polygon using its vertices using
4      Green's theorem.
5      inputs: vertices - List of tuples/lists, each contains x and y
6                          coordinates of a vertex of the polygon, eg:
7                          [(-1, -1), (1, -1), (3, -1), (1, 1), (-1, 1)]
8      return: Area of the polygon.
9      Code suggested by ChatGPT.
10     """
11     n = len(vertices)
12     area = 0.0
13
14     for i in range(n):
15         x1, y1 = vertices[i]
16         x2, y2 = vertices[(i+1)%n]   # Next vertex, and
17                                      # wrapping around to the 1st vertex
18         area += (x1*y2 - y1*x2)
19
20     return area/2
```

4.2.4 *The dot-product formula*

Equation (4.19) can be expressed as two separate summation terms:

$$A = \frac{1}{2} \sum_{i=1}^{n} (x_i y_{i+1} - y_i x_{i+1}) = \frac{1}{2} \sum_{i=1}^{n} x_i y_{i+1} - \frac{1}{2} \sum_{i=1}^{n} y_i x_{i+1} \qquad (4.20)$$

We can now observe a clear pattern in Eq. (4.20): The first term represents the dot product of the vector $[x_1, x_2, \ldots, x_n]^\top$ with the vector $[y_2, y_3, \ldots, y_n, y_1]^\top$ (shifted one position backward). Similarly, the second term is the dot product of the vector $[y_1, y_2, \ldots, y_n]^\top$ with the vector

$[x_2, x_3, \ldots, x_n, x_1]^{\top}$ (also shifted one position backward). This computation can be efficiently expressed using the following formula:

$$A_{\text{polygon}} = \left(\mathbf{x} \cdot \text{roll}(\mathbf{y}, -1) - \mathbf{y} \cdot \text{roll}(\mathbf{x}, -1)\right)/2. \tag{4.21}$$

Here, $\text{roll}(\mathbf{v}, n)$ is a built-in function commonly used in modern programming languages like Python. It performs a cyclic permutation of the elements in the array $\mathbf{v}$ by n positions. For instance, for an array $\mathbf{v} = (1, 2, 3, 4)$, $\text{roll}(\mathbf{v}, -1)$ shifts the elements left by 1, resulting in a new vector $\mathbf{v} = (2, 3, 4, 1)$. This operation aligns precisely with the operation needed in Eq. (4.20). The dot-product operation can often be optimized for parallel processing, leading to faster computation.

This dot-product formula was derived for quadrilateral domains (Q4) in Ref. [4]. When the author observed the pattern in the formula for Q4, they anticipated that the formula should also work for general polygons with an arbitrary number of edges. This hypothesis was tested with examples for T3 and P5 in Ref. [4], though without proof. Now, Eq. (4.20) serves as the proof. Note that **gr.polygonA_xy()** produces the same result as **polygon_area_green()**, as will be demonstrated in the following example. The data formats are slightly different and require data conversion. The following code illustrates that these two formulas yield the same result:

```
1  # Test for Q4 polygons with 4 nodes:
2
3  n = 4
4  xn=sp.symbols(f'x:{n}')                    # for polygons with n edges
5  yn=sp.symbols(f'y:{n}')
6
7  A_roll  = (xn@np.roll(yn,-1)-yn@np.roll(xn,-1))/2
8  display(Math(f" \\text{{A_roll}}  = {latex(A_roll)}"))
9
10 A_green = polygon_area_green([(xi, yi) for xi, yi in zip(xn,yn)])
11 display(Math(f" \\text{{A_green}} = {latex(A_green)}"))
```

$$A_{\text{roll}} = \frac{x_0 y_1}{2} - \frac{x_0 y_3}{2} - \frac{x_1 y_0}{2} + \frac{x_1 y_2}{2} - \frac{x_2 y_1}{2} + \frac{x_2 y_3}{2} + \frac{x_3 y_0}{2} - \frac{x_3 y_2}{2}$$

$$A_{\text{green}} = \frac{x_0 y_1}{2} - \frac{x_0 y_3}{2} - \frac{x_1 y_0}{2} + \frac{x_1 y_2}{2} - \frac{x_2 y_1}{2} + \frac{x_2 y_3}{2} + \frac{x_3 y_0}{2} - \frac{x_3 y_2}{2}$$

The equivalence is clear.

4.2.5 *Examples: Areas of polygons*

The following code computes the area of a triangle using its vertices
numbered counterclockwise:

```
1  T3_vertices = [(0, 0), (1, 0), (0, 1)]      # 3 vertices for a triangle
2  A_T3 = polygon_area_green(T3_vertices)
3  print(f"The area of the triangle is: {A_T3}")
4
5  x_=[]; y_=[]
6  for xi, yi in T3_vertices:
7      x_.append(xi); y_.append(yi)
8
9  A_T3_ = gr.polygonA_xy(x_, y_)
10 print(f"The area obtained using gr.polygonA_xy(): {A_T3_}")
```

```
The area of the triangle is: 0.5
The area obtained using gr.polygonA_xy(): 0.5
```

The following code computes the area of a pentagon with five nodes
(P5) using vertices numbered in a counterclockwise direction. The nodal
coordinates are given in the following format:

```
1  # coordinates for a P5, in x-y coordinates:
2
3  # Nodes:        1  2  3  4  5
4  x_CCW=np.array([-1, 1, 3, 1,-1])
5  y_CCW=np.array([-1,-1,-1, 1, 1])
6
7  P5_vertices=[(xi,yi) for xi,yi in zip(x_CCW, y_CCW)] # data conversion
8  P5_vertices
```

```
[(-1, -1), (1, -1), (3, -1), (1, 1), (-1, 1)]
```

```
1  # Compute the area:
2
3  A_P5 = polygon_area_green(P5_vertices)
4  print(f"The area of the pentagon is: {A_P5}")
5
6  A_P5_ = gr.polygonA_xy(x_CCW, y_CCW)
7  print(f"The area obtained using gr.polygonA_xy(): {A_P5_}")
```

```
The area of the pentagon is: 6.0
The area obtained using gr.polygonA_xy(): 6.0
```

We obtained the same results. Note that **polygonA_xy()**, developed in
Chapter 5 of Ref. [4], can be more efficient than **polygon_area_green()**
when the number of edges is large. This efficiency arises because we utilized
the built-in **roll()** function, allowing the computation to be performed as
a dot product.

Additionally, Eq. (4.19) can naturally be expressed in the same form using the roll() function owing to the closed curve integration in Green's theorem and the established sign convention. Both formulas are precisely the same.

4.3 A proof for conservative fields using Green's theorem

As discussed in Chapter 3, a conservative field has a continuously differentiable potential function $\phi(x, y)$. The components of the force vector $\mathbf{F}$ can be expressed as $F_x = \frac{\partial \phi}{\partial x}$ and $F_y = \frac{\partial \phi}{\partial y}$. By substituting these expressions into Eq. (4.1), we have

$$\underbrace{\oint_{\mathcal{C}} (F_x \, dx + F_y \, dy)}_{\text{total work done along the curve } \mathcal{C}} = \underbrace{\iint_{\mathcal{D}} \left(\frac{\partial^2 \phi}{\partial y \partial x} - \frac{\partial^2 \phi}{\partial x \partial y} \right) dx \, dy = 0}_{\text{total circulation in } \mathcal{D}} \qquad (4.22)$$

This is because $\phi(x, y)$ is continuously differentiable, and hence the mixed partial derivatives are order-independent: $\frac{\partial^2 \phi}{\partial y \partial x} = \frac{\partial^2 \phi}{\partial x \partial y}$. This implies that the work done by any conservative force field along a closed curve must be zero. From Eq. (4.22), the following is also clear:

> A conservative field has a zero curl.

On the left side of Eq. (4.22), the work done is given as $W = \phi(b) - \phi(a)$ for any a and b. When the curve is closed, $a = b$, so we have $\phi(b) - \phi(a) = 0$. Since ϕ is differentiable, implying continuity, we have $W = 0$.

We have seen such examples earlier. Green's theorem puts all these pieces together. This also implies that Green's theorem is useful for fields that are nonconservative, where the integral values are not zero but equal. The domain integral measures the curl component of a vector field. It provides the option to evaluate the integrals using either the curve integral or the 2D domain integral.

4.4 Green's theorem across boundary and over domain

Green's theorem also holds for curve integrals across a boundary and domain integrals over a domain. For ease of discussion, let us call this **Green's divergence theorem.**

Consider a differentiable vector field defined in the domain $\mathcal{D}$ and denoted as

$$\mathbf{V} = \begin{bmatrix} M(x, y) \\ N(x, y) \end{bmatrix} \qquad (4.23)$$

At the boundary $\mathcal{C}$ that bounds $\mathcal{D}$, the media can flow in and out. Green's divergence theorem for curve integrals across the boundary of a domain and over the domain is expressed as

$$\oint_{\mathcal{C}} (M \, dy - N \, dx) = \iint_{\mathcal{D}} \left(\frac{\partial M}{\partial x} + \frac{\partial N}{\partial y} \right) dx \, dy \tag{4.24}$$

Comparing Eq. (4.1) and Eq. (4.24), it is seen that these two equations are mathematically identical. If we replace $-M$ and N in Eq. (4.1) with N and M, respectively, we obtain Eq. (4.24). Therefore, the proof of Eq. (4.24) is essentially the same as that of Eq. (4.1). Hence, we will not repeat it here.

The major difference lies in the application. The closed curve integral in Eq. (4.24) measures the media flow across the boundary $\mathcal{C}$. Therefore, the direction of interest for the flow is the normal direction $\mathbf{n}$, which is rotated $90°$ clockwise from the tangent direction $\mathbf{t}$, as discussed in the previous chapter.

Let us examine the physical meaning of the term on the right side of Eq. (4.24) by considering a flowing medium, such as water, with a velocity field: $M(x, y) = V_x(x, y)$ and $N(x, y) = V_y(x, y)$.

4.4.1 *Divergence of a vector field*

Consider an arbitrary vector field $\mathbf{V}$ defined in Eq. (4.23) that is differentiable in domain $\mathcal{D}$.

Let us examine an infinitely small rectangular cell $dx \, dy$ in $\mathcal{D}$, as shown in Fig. 4.5.

Over the left edge of the cell, the field strength is $V_x \, dy$. Over the right edge of the cell, the field strength is $(V_x + dV_x) \, dy$, where dV_x is due to the increment in dx. Therefore, we have

$$dV_x = \frac{\partial V_x}{\partial x} dx \tag{4.25}$$

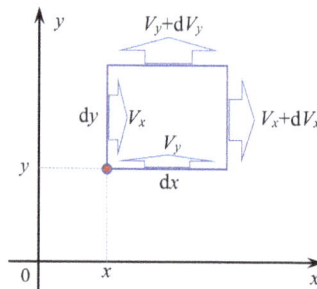

Figure 4.5. An infinitely small rectangular cell $dx \, dy$ in the domain $\mathcal{D}$.

This is because $V_x(x, y)$ is assumed to be differentiable. Over the bottom edge of the cell, the field strength is $V_y \, dx$, and over the top edge of the cell, it is $(V_y + dV_y) \, dx$, where dV_y is due to dy, and hence it should be

$$dV_y = \frac{\partial V_y}{\partial y} dy \tag{4.26}$$

The net change in the strength of the field vector in the x-direction becomes

$$dV_x \, dy = \frac{\partial V_x}{\partial x} dx \, dy \tag{4.27}$$

For the same reason, the net change in the strength of field vector in the y-direction is

$$dV_y \, dx = \frac{\partial V_y}{\partial y} dy \, dx \tag{4.28}$$

The total field strength change in the cell becomes

$$dV_x \, dy + dV_y \, dx = \underbrace{\left(\frac{\partial V_x}{\partial x} + \frac{\partial V_y}{\partial y} \right)}_{\text{divergence of vector } \mathbf{V}} dy \, dx \tag{4.29}$$

where

$$\text{div} V = \frac{\partial V_x}{\partial x} + \frac{\partial V_y}{\partial y} \tag{4.30}$$

This is called the **divergence** of vector $\mathbf{V}$ in the infinitely small cell at point (x, y) in the domain. It is a measure of the distributed source at point (x, y) within $\mathcal{D}$. Therefore, the integral on the right side of Eq. (4.24) gives the total divergence of the velocity field in domain $\mathcal{D}$. It becomes a scalar that matches the scalar value on the left side of Eq. (4.24). This represents the net change in the amount of fluid at a point. After the integration, we obtain the total divergence of vector $\mathbf{V}$ in $\mathcal{D}$:

$$\iint_{\mathcal{D}} \left(\frac{\partial V_x}{\partial x} + \frac{\partial V_y}{\partial y} \right) dx \, dy \Rightarrow \quad \text{total divergence of vector } \mathbf{V} \text{ in } \mathcal{D} \tag{4.31}$$

Therefore, Eq. (4.24) is a mathematical description of the **conservation law** of mass:

The total flux across the closed boundary of a domain equals the total divergence of the field over the domain.

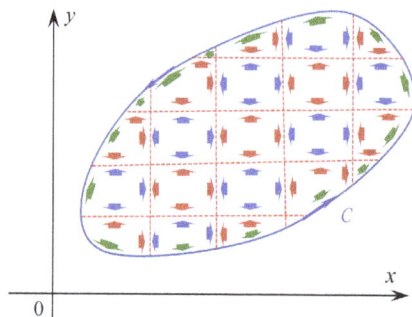

Figure 4.6. Cancellation of the divergence of a vector function within the integral domain, leaving an integral of the vector function along the boundary curve.

4.4.2 *A pictorial proof of Green's theorem, divergence*

This section provides a pictorial description to explicitly illustrate the cancellation of the divergence of a vector function within the integral domain, leaving only an integral of the vector function along the boundary curve.

Figure 4.6 shows a discretized domain $\mathcal{D}$ bounded by $\mathcal{C}$, with very small cells. The arrows represent the divergence along each of the cell edges. It is clear that all the red and blue arrows on the same edge will cancel each other out, as each pair of red and blue arrows has exactly the same value (even though we may not know the value) but in opposite directions. This cancellation occurs for all the internal edges, leaving only the green arrows along the edges on $\mathcal{C}$, as shown in Fig. 4.6.

The rigorous proof is quite similar to that given in Section 4.1, so it will not be repeated here. It is a useful exercise for readers to work out the detailed proof, carefully following the rules of directions for both domain and curve integrals.

4.4.3 *The first fundamental theorem of calculus in 2D*

Readers may compare the divergence theorem with the **first fundamental theorem of calculus** that we studied in Section 4.1.7 of Ref. [4]. This comparison provides insight into how divergence in the 1D case is summed up and why it equals the flux only on the boundary. The divergence theorem can be seen as a dimensional extension of the first fundamental theorem of calculus, which is based on 1D integration.

4.4.4 *Python code for examination of Green's theorem*

To efficiently examine Green's theorem for different vector fields, we provide the following Python function. The tasks that this function aims to perform are listed in its docstring:

```python
def exam_GreensTheorem2(V, X, a):
    '''Examine a given vector field V that is differentiable over
        a circular domain. The code prints out:
        1 The divergence of the vector field V.
        2 The total divergence of the field over a circular domain.
        3 The total flux of V across the boundary curve of the circle.
        4 Whether or not the Green's divergence theorem holds.
        The size of of the domain is controlled by the radius a.
    '''

    r, θ = symbols('r, θ', real=True) # for integration in polar coord.
    x, y = X                          # unpack the Cartesian coordinates

    # Coordinate transformation, for integral over the closed circle C:
    xC = a*cos(θ); yC = a*sin(θ)           # xC, yC are x,y on curve C
    dicC = {x:xC, y:yC}

    nds = Matrix([yC.diff(θ), -xC.diff(θ)])
    fIC = V.T@nds                           # integrand for the curve integral
    fIC = fIC.subs(dicC)
    display(Math(f"\\text{{ Integrand of the curve integral}}=\
                                {latex(sp.simplify(fIC)[0])}"))
    Int = integrate
    Vns = Int(fIC, (θ,0,2*sp.pi)).simplify()[0]

    # Integral over the circle D:
    xD = r*cos(θ)                    # xD, yD are x,y in the circular Domain
    yD = r*sin(θ)
    dicD = {x:xD, y:yD}
    div_V = gr.div_vf(V, X).simplify()   # the divergence in the Domain
    display(Math(f" \\text{{ Divergence of the field}}={latex(div_V)}"))

    fID = (div_V*r).simplify() # integrand in circular domain, detJ=r
    divVD = Int(Int(fID.subs(dicD),(θ,0,2*sp.pi)),(r,0,a)).simplify()

    display(Math(f" \\text{{Total flux of the vector field across the \
                            circular boundary}} = {latex(Vns)}"))
    display(Math(f" \\text{{Total divergence over the circle}} \
                            = {latex(divVD)}"))
    print(f" Does Green's divergence theorem hold?{Vns==divVD}")
```

4.4.5 *Example: Green's theorem on divergence-free spin fields*

Consider the spin field $\mathbf{s}_z = [-y\ \ x]^\top$ studied earlier. Compute the following:

1. the divergence of the field,
2. the total flux of the field across the boundary of a circular domain centered at the origin,
3. the total divergence of the field in the circular domain,
4. check whether Green's theorem holds.

Solution: We use exam_GreensTheorem2() and the following code snippet to perform these tasks:

```
1  x, y = symbols('x, y', real=True)
2  a = symbols('a', positive=True)          # radius of the circle
3  X = Matrix([x, y])
4
5  sz = Matrix([-y, x])                       # the given vector field
6  exam_GreensTheorem2(sz, X, a)
```

Integrand of the curve integral $= 0$

Divergence of the field $= 0$

Total flux of the vector field across the circular boundary $= 0$

Total divergence over the circle $= 0$

Does Green's divergence theorem hold? True

As observed, the divergence of the spin field is zero, as found in the previous chapter. The flux is also zero because the spin field has zero partial derivatives for its component functions across the entire domain. Green's theorem holds as $0 = 0$.

4.4.6 *Example: Position field with constant divergence*

Consider the same position fields given in Eq. (4.10).

Show the following:

1. These fields have a constant divergence.
2. The curve integral of flux over a circular boundary centered at the origin equals the integral of the divergence over the domain with the circular boundary, and Green's theorem holds.

Solution: We use the same Python function and the following snippet to perform these tasks:

```
1  # Consider r1:
2  r1 = Matrix([x, y ])                        # the given vector field
3  exam_GreensTheorem2(r1, X, a)
```

Integrand of the curve integral $= a^2$

Divergence of the field $= 2$

Total flux of the vector field across the circular boundary $= 2\pi a^2$

Total divergence over the circle $= 2\pi a^2$

Does Green's divergence theorem hold? True

```
1  # Consider r2:
2  r2 = Matrix([x, y+2*a ])
3  exam_GreensTheorem2(r2, X, a)
```

Integrand of the curve integral $= a^2 \cdot (2\sin(\theta) + 1)$

Divergence of the field $= 2$

Total flux of the vector field across the circular boundary $= 2\pi a^2$

Total divergence over the circle $= 2\pi a^2$

Does Green's divergence theorem hold? True

We found the following:

1. The divergence of the 2D position fields (radial flow) is 2.
2. The total divergence over a circle is two times the area of the circle. This result can also be used to compute the area of a domain, as discussed earlier.
3. The flux across the closed circle equals the total divergence in the domain, and Green's theorem holds.

These findings are also true for r_2, implying that the center of the radial flow is immaterial in these cases. This is because the construction of the position field yields the same constant divergence of 2 throughout the entire space. The total divergence of any domain is two times the area of the domain regardless of its shape.

4.4.7 *Example: Green's theorem on a singular position field*

Consider the following position field:

$$\mathbf{r}_{r2} = \begin{bmatrix} \frac{x}{r^2} & \frac{y}{r^2} \end{bmatrix}^\top \tag{4.32}$$

where $r = \sqrt{x^2 + y^2}$.

Perform the following **tasks**:

1. Compute the divergence of the field.
2. Compute the flux of the field across the circular boundary of a circular domain centered at the origin.
3. Compute the domain integral of the divergence over the circular domain.
4. Check whether Green's divergence theorem holds.

Solution for the field $\mathbf{r}_{r2}$: This is the position field **r** divided by r^2, which has a singular point at the origin. We use `exam_GreensTheorem2()` and the following code snippet to complete the tasks:

```
1  rr2x = x/(x**2+y**2)
2  rr2y = y/(x**2+y**2)
3  V = Matrix([rr2x, rr2y])
4  exam_GreensTheorem2(V, X, a)
```

Integrand of the curve integral $= 1$

Divergence of the field $= 0$

Total flux of the vector field across the circular boundary $= 2\pi$

Total divergence over the circle $= 0$

Does Green's divergence theorem hold? False

This time, we found that Green's divergence theorem does not hold. The reason is similar to the case of the field $\mathbf{s}_{r2} = \begin{bmatrix} -\frac{y}{r^2} & \frac{x}{r^2} \end{bmatrix}^\top$ studied earlier. Both fields have a singularity at $(0,0)$. The "False" assessment is misleading as the theorem is not supposed to hold for nondifferentiable vector functions. If the circle is placed anywhere in the domain, excluding the origin, Green's divergence theorem will hold true.

4.4.8 *Example: Green's theorem on the 2D gravity field*

Consider the gravity field studied in Section 3.6.2. Here, we use a circular domain and its boundary curve and conduct the same examination as in the previous example using the same code:

```
1  m, g = symbols('m, g', positive=True)           # mass and gravity
2
3  GFv = sp.Matrix([0*g, -m*g])                     # vertical gravity field
4  exam_GreensTheorem2(GFv, X, a)
```

Integrand of the curve integral $= -agm \sin(\theta)$

Divergence of the field $= 0$

Total flux of the vector field across the circular boundary $= 0$

Total divergence over the circle $= 0$

Does Green's divergence theorem hold? True

We found zero divergence because the gravity field is a constant field, and the divergence is always zero throughout the entire 2D space, resulting in a zero integrand for the domain integration part. The curve integral is also zero (although the integrand is nonzero) because the curve is enclosed, and the gradient field is conservative, as discussed in Chapter 3. In any case, Green's theorem holds.

4.4.9 *Example: Green's theorem for a mixed field*

Consider a field that is a mixture of the position and spin fields defined in Eq. (4.12).

The snippet to examine it is as follows:

```
1  Vx, Vy = symbols('V_x, V_y', real=True)
2  Vx = x-y; Vy = y+x                               # a mixed field
3  V = Matrix([Vx, Vy])
4  exam_GreensTheorem2(V, X, a)
```

Integrand of the curve integral $= a^2$

Divergence of the field $= 2$

Total flux of the vector field across the circular boundary $= 2\pi a^2$

Total divergence over the circle $= 2\pi a^2$

Does Green's divergence theorem hold? True

It is observed that these integrals capture the divergence part of the field and ignore the circulation part.

4.4.10 *Example: Green's theorem for an arbitrary field*

Consider the arbitrarily constructed vector field given in Eq. (4.13).

The snippet to examine it is as follows:

```
1  Vx, Vy = symbols('V_x, V_y', real=True)
2  Vx = -x**2*y+x; Vy = x**2                    # an arbitrary field
3  V = Matrix([Vx, Vy])
4  exam_GreensTheorem2(V, X, a)
```

Integrand of the curve integral $= a^2 \left(-\dfrac{a^2 \sin{(2\theta)}}{2} + a \sin{(\theta)} + 1 \right) \cos^2{(\theta)}$

Divergence of the field $= -2xy + 1$

Total flux of the vector field across the circular boundary $= \pi a^2$

Total divergence over the circle $= \pi a^2$

Does Green's divergence theorem hold? True

The nonlinear vector field has a divergence that varies within the domain. Green's theorem also holds. The curve integral captures the flux of the field across the circular boundary, while the domain integral captures the total divergence of the field. These are equal.

In summary, we note the following:

Both theorems hold for differentiable vector fields.

Green's circulation theorem captures the circulation of a vector field in a domain, which corresponds to its work done along the boundary of the domain.

Green's divergence theorem captures the divergence of a vector field in a domain, which corresponds to its flux across the boundary of the domain.

Let us summarize the major differences between the two Green's theorem expressions.

4.5 Differences between two Green's theorem expressions

The differences on both sides of these two Green's theorem expressions are as follows:

$$
\begin{array}{ccc}
\text{work done along } \mathcal{C} & \text{versus} & \text{flux across } \mathcal{C} \\[2mm]
\displaystyle\oint_{\mathcal{C}} (M \, \mathrm{d}x + N \, \mathrm{d}y) & \text{versus} & \displaystyle\oint_{\mathcal{C}} (M \, \mathrm{d}y - N \, \mathrm{d}x) \\[2mm]
\text{curl over } \mathcal{D} & \text{versus} & \text{divergence over } \mathcal{D} \\[2mm]
\displaystyle\iint_{\mathcal{D}} \left(\frac{\partial N}{\partial x} - \frac{\partial M}{\partial y} \right) \mathrm{d}x \, \mathrm{d}y & \text{versus} & \displaystyle\iint_{\mathcal{D}} \left(\frac{\partial M}{\partial x} + \frac{\partial N}{\partial y} \right) \mathrm{d}x \, \mathrm{d}y
\end{array}
\tag{4.33}
$$

For a **force field**, we have

$$\oint_C (F_x \, dx + F_y \, dy) = \iint_D \left(\frac{\partial F_y}{\partial x} - \frac{\partial F_x}{\partial y} \right) dx \, dy \qquad (4.34)$$

$$\underbrace{}_{\text{total work along } C} \qquad \underbrace{\phantom{\iint_D \left(\frac{\partial F_y}{\partial x} - \frac{\partial F_x}{\partial y} \right) dx \, dy}}_{\text{total circulation in } D}$$

or simply

$$\oint_C \mathbf{V} \cdot \mathbf{t} ds = \iint_D \text{crul } \mathbf{V} dA \qquad (4.35)$$

where crul$\mathbf{V}$ is the 2D curl, which is the third components of the 3D curl.

For a **velocity field**, we have:

$$\oint_C (-V_y \, dx + V_x \, dy) = \iint_D \left(\frac{\partial V_x}{\partial x} + \frac{\partial V_y}{\partial y} \right) dx \, dy \qquad (4.36)$$

$$\underbrace{}_{\text{total flux across } C} \qquad \underbrace{\phantom{\iint_D \left(\frac{\partial V_x}{\partial x} + \frac{\partial V_y}{\partial y} \right) dx \, dy}}_{\text{total divergence in } D}$$

or simply

$$\oint_C \mathbf{V} \cdot \mathbf{n} ds = \iint_D \nabla \cdot \mathbf{V} dA \qquad (4.37)$$

These Green's theorem expressions are all conservative laws but apply to different quantities and vector fields.

Due to these differences, the characteristics of a conservative field and a divergence-free field are generally opposites. For example, the position field (radial, irrotational) is conservative but not divergence-free. The spin field (rotational) is divergence-free but nonconservative.

4.6 Divergence-free vector fields

There are some special and important fields. The conservative field is one of these, and we studied it in Chapter 3. Another important type is divergence-free fields.

4.6.1 *Derivation of stream functions*

As the name suggests, a divergence-free field is a vector field with zero divergence at any point in the domain:

$$\nabla \cdot \mathbf{V} = \frac{\partial V_x}{\partial x} + \frac{\partial V_y}{\partial y} + \frac{\partial V_z}{\partial z} = 0 \qquad (4.38)$$

The condition for a vector field $\mathbf{V}$ to be divergence-free is that there exists a continuously differentiable **vector potential function** denoted as

$$\mathbf{A} = [A_x \ A_y \ A_z]^\top \tag{4.39}$$

such that

$$\mathbf{V} = \nabla \times \mathbf{A} = \begin{bmatrix} \frac{\partial A_z}{\partial y} - \frac{\partial A_y}{\partial z} \\ \frac{\partial A_x}{\partial z} - \frac{\partial A_z}{\partial x} \\ \frac{\partial A_y}{\partial x} - \frac{\partial A_x}{\partial y} \end{bmatrix} \tag{4.40}$$

Here, we used Eq. (2.21). Clearly, if such an $\mathbf{A}$ can be found, we automatically satisfy the divergence-free condition:

$$\nabla \cdot \mathbf{V} = \nabla \cdot (\nabla \times \mathbf{A}) = 0 \tag{4.41}$$

where we used the identity given in Eq. (2.29).

In contrast to a potential field, which has a potential function, a divergence-free field has a stream function.

Consider a simpler case where the vector field describes a two-dimensional flow in the x–y plane. In this case, all functions involved do not depend on the z-axis (although the field may have a z-component). The vector field can be expressed as

$$\mathbf{V} = \begin{bmatrix} V_x(x, y) \\ V_y(x, y) \\ c \end{bmatrix} \tag{4.42}$$

where c is a constant. Since all functions, including the components of the vector potential, do not change with z, Eq. (4.40) is reduced to

$$\mathbf{V} = \begin{bmatrix} \frac{\partial A_z}{\partial y} \\ -\frac{\partial A_z}{\partial x} \\ \frac{\partial A_y}{\partial x} - \frac{\partial A_x}{\partial y} \end{bmatrix} \tag{4.43}$$

Comparing Eqs. (4.42) and (4.43), we have

$$V_x = \frac{\partial A_z}{\partial y}; \quad V_y = -\frac{\partial A_z}{\partial x} \tag{4.44}$$

where A_z is called the **stream function** for 2D divergence-free vector fields, whose partial derivatives yield the component functions of the vector field.

It is, in fact, the third component of the vector potential field in 3D cases. Conventionally (in fluid mechanics), it is often denoted as $\psi(x, y)$:

$$V_x = \frac{\partial \psi}{\partial y}; \quad V_y = -\frac{\partial \psi}{\partial x} \tag{4.45}$$

Finally, it is easy to verify that the vector field $\mathbf{V}$ with components given in Eq. (4.45) satisfies (the 2D case of) Eq. (4.38):

$$\nabla \cdot \mathbf{V} = \frac{\partial V_x}{\partial x} + \frac{\partial V_y}{\partial y} = \frac{\partial^2 \psi}{\partial x \partial y} - \frac{\partial^2 \psi}{\partial y \partial x} = 0 \tag{4.46}$$

Here, we used the continuously differentiable nature of the stream function.

Note that for a 2D vector field defined in Eq. (4.42), the first two components of the vector potential should satisfy

$$\frac{\partial A_y}{\partial x} - \frac{\partial A_x}{\partial y} = c \tag{4.47}$$

if we need to determine these components.

4.6.2 *Streamlines*

The reason for calling $\psi(x, y)$ a stream function is that it characterizes the streamlines in a fluid flow. A streamline is defined as a line in a flow field that is always tangential to the velocity vector at any point in the field, as shown in Fig. 4.7. These lines can be observed by dropping some ink into the flow. The trace lines of the ink are the streamlines.

Consider now a differential line segment vector $\mathbf{dr} = [dx \ dy]^\top$ at a point (x, y) on a streamline in the domain. Using the definition of the streamline, we have

$$\mathbf{V} \times \mathbf{dr} = 0 \tag{4.48}$$

due to the direction alignment of $\mathbf{V}$ and $\mathbf{dr}$, as shown in Fig. 4.8.

Equation (4.48) gives

$$\mathbf{V} \times \mathbf{dr} = V_x dy - V_y dx = 0 \tag{4.49}$$

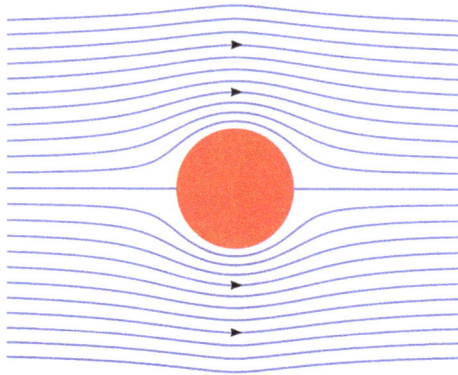

Figure 4.7. Streamlines for the incompressible potential flow around a circular cylinder. These lines are always tangential to the flow velocity at any point in the domain. The stream function value is constant along a streamline.
Source: Image from Wikimedia Commons under the CC BY-SA 3.0 license and those by Kraaiennest.

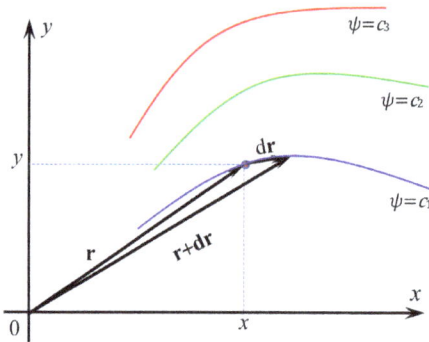

Figure 4.8. A differential segment vector on a streamline. Streamlines form a contour of the stream function. Three streamlines are shown for a 2D function ψ with $\psi(x, y) = c_1$, $\psi(x, y) = c_2$, and $\psi(x, y) = c_3$, where $c_i (i = 1, 2, \ldots)$ are constants.

Here, we used the cross-product formula for two vectors in 2D (Eq. (2.12)).

Substituting Eq. (4.45) into Eq.(4.49), we have

$$\frac{\partial \psi}{\partial y} dy + \frac{\partial \psi}{\partial x} dx = d\psi = 0 \qquad (4.50)$$

This means that $\psi(x, y)$ is constant along a streamline; hence, it is called the stream function. These streamlines form the contour lines of the stream function. Three such lines are schematically shown in Fig. 4.8.

4.6.3 *Properties of a divergence-free field*

We have seen that if the vector field is constant, such as the 2D gravity field, its divergence is zero. There are many nonconstant vector fields that have zero divergence. The properties of a divergence-free field are as follows:

1. The net flux $\oint_C \mathbf{V} \cdot \mathbf{n} \, ds$ across any closed curve is zero.
2. The flux $\int_a^b \mathbf{V} \cdot \mathbf{n} \, ds$ depends only on points a and b. It does not depend on the path chosen for the integration from a to b.
3. There exists a continuously differentiable stream function $\psi(x, y)$, whose partial derivatives give the component functions of the vector field:

$$V_x = \frac{\partial \psi}{\partial y}; \quad V_y = -\frac{\partial \psi}{\partial x} \tag{4.51}$$

4. Zero divergence at any point in the domain:

$$\frac{\partial V_x}{\partial x} + \frac{\partial V_y}{\partial y} = 0 \tag{4.52}$$

A divergence-free field that possesses one of these properties has all of them.

Note that the constant in a stream function ψ is immaterial in terms of generating the components of the field because the component functions are obtained using its partial derivatives. Therefore, we often need to find only the nonconstant part of the potential function ψ if it exists for a given vector field. Once it is found, adding a constant gives an additional streamline (or contour line).

4.6.4 *Example: The spin field and its stream function*

This example proves that the nonconservative spin field $[-y \ x]^T$ is divergence-free and hence source-free. We then proceed to find its stream function.

First, we use the following code to show that item 4 can be satisfied:

```
1  x, y = symbols('x, y', real=True)              # define variables
2  Sx = -y; Sy =  x                                # spin field
3  # Check item 4:
4  display(Math(f" \\frac{{∂V_x}}{{∂x}}+\\frac{{∂V_y}}{{∂y}}=0?\;\; \
5                 {latex((Sx.diff(x)+Sy.diff(y)).simplify()==0)}"))
6  print(f"Equations to use to find ψ: ∂ψ/∂x = {-Sy}; ∂ψ/∂y = {Sx}")
```

$$\frac{\partial V_x}{\partial x} + \frac{\partial V_y}{\partial y} = 0? \text{ True}$$

```
Equations to use to find ψ: ∂ψ/∂x = -x; ∂ψ/∂y = -y
```

Item 4 is satisfied. Therefore, we can find the stream function for it:

```
1  # First, use condition: ∂ψ/∂x = -Sy to find possible ψ:
2
3  C = sp.Function('C')(y)        # define symbolic integral function
4  ψC = integrate(-Sy, (x)) + C         # add an unknown integral function
5  ψC                                    # possible stream function
```

$$-\frac{x^2}{2} + C(y)$$

```
1  # Next, use condition: ∂ψ/∂y = Sx=-y to find C(y):
2
3  ψC.diff(y)
```

$$\frac{d}{dy}C(y)$$

Since $\frac{\partial C(y)}{\partial y} = -y$, we have $C(y) = -\frac{1}{2}y^2$. Now, substituting $C(y)$ to the expression of ψC, we obtain the following:

```
1  ψ = ψC.subs(C, -y**2/2)
2  display(Math(f" \\text{{The the stream function found = }}{latex(ψ)}"))
```

The stream function found $= -\dfrac{x^2}{2} - \dfrac{y^2}{2}$

Lastly, we check whether the potential function found satisfies the conditions in item 1:

```
1  print(f"Is -Sy = ψ.diff(x)? {-Sy == ψ.diff(x)}")
2  print(f"Is Sx = ψ.diff(y)? {Sx == ψ.diff(y)}")
```

```
Is -Sy = ψ.diff(x)? True
Is Sx = ψ.diff(y)? True
```

Finally, the stream function for the spin field can be expressed as

$$\psi(x, y) = -\frac{1}{2}(x^2 + y^2) \tag{4.53}$$

4.6.5 *Example: The position field, not divergence-free*

Let us show that the conservative position field $\mathbf{r} = [x \quad y]^\top$ is not divergence-free and a stream function cannot be found:

```
1  x, y = symbols('x, y', real=True)                    # define variables
2  rx = x; ry = y                                        # the position field
3  # Check item 4:
4  display(Math(f" \\frac{{∂V_x}}{{∂x}}+\\frac{{∂V_y}}{{∂y}}=0?\;\; \
5                     {latex((rx.diff(x)+ry.diff(y)).simplify()==0)}"))
6  print(f"Equations to use to find ψ: ∂ψ/∂x = {-ry}; ∂ψ/∂y = {rx}")
```

$$\frac{\partial V_x}{\partial x} + \frac{\partial V_y}{\partial y} = 0? \ \text{False}$$

```
Equations to use to find ψ: ∂ψ/∂x = -y; ∂ψ/∂y = x
```

```
1  # Use ∂ψ/∂x =-y:
2
3  C = sp.Function('C')(y)              # define symbolic integral function
4  ψC = integrate(-y, x) + C
5  ψC                                   # possible stream function
```

$$-xy + C(y)$$

```
1  # Use ∂ψ/∂y = x:
2
3  D = sp.Function('D')(x)
4  ψD = integrate(x, y) + D
5  ψD                                   # possible stream function
```

$$xy + D(x)$$

Since $C(y)$ is only a function of y and $D(x)$ is only a function of x, it is not possible to find such $C(y)$ and $D(x)$ to make ψ_C equal to ψ_D.

Since the position field is not divergence-free, the flux generated by it will be path-dependent.

As seen in these examples, the procedure for analyzing, testing, and finding stream functions is quite similar to that for potential functions except for the key formulation differences. Without repeating too many detailed discussions, we provide the code for systematically finding stream functions for divergence-free vector fields in the following section.

4.6.6 *Python code for finding stream functions*

```
1  def find_stream_f2D(Vx, Vy, X):
2      '''
3          Find the stream function ψ for a given set of vector field
4          components. The vector field must be a divergence free field.
5          return: ψ - the stream function found.
6      '''
7      x, y = X
8      # Use condition: ∂ψ/∂x =-Vy
9      C = sp.Function('C')(y)          # define symbolic integral functions
10     ψC = integrate(-Vy, (x)) + C     # ψ with C(y); used -Vy = ∂ψ/∂x
11
12     # use condition: Vx = ∂ψ/∂y
13     eqn = ψC.diff(y)- Vx                 # create an equation to solve
14     sln_C = sp.dsolve(eqn, C)            # solve the eqn for C(y)
15
16     ψC = ψC.subs(C, sln_C.args[1])   # put solution of C(y) back to ψC
17     ψ  = ψC.subs(ψC.args[0],0).simplify() # set integral constant to 0
18     gr.printx('ψ')
19
20     # Check all conditions:
21     gr.printx('(ψ.diff(x)+Vy).simplify()')          # check condition
22     gr.printx('(ψ.diff(y)-Vx).simplify()')          # check condition
23
24     return ψ
```

Let's test it for the $\mathbf{s}_z$ field studied previously:

```
1  ψ = find_stream_f2D(Sx, Sy, X)
2  gr.printM(ψ, ' The stream function ψ found is: ')
```

```
ψ = -x**2/2 - y**2/2
(ψ.diff(x)+Vy).simplify() = 0
(ψ.diff(y)-Vx).simplify() = 0
 The stream function ψ found is:
```

$$-\frac{x^2}{2} - \frac{y^2}{2}$$

The stream function found is the same as the one we obtained earlier. Readers may test other cases.

Caution is needed when using **find_stream_f2D()**. If the field is not divergence-free, the code may still produce a function; however, this output will fail the divergence-free tests. For example, for the position field that has been proven not to be divergence-free, the results are as follows:

```
1  ψ = find_stream_f2D(rx, ry, X)
```

ψ = x*y
(ψ.diff(x)+Vy).simplify() = 2*y
(ψ.diff(y)-Vx).simplify() = 0

As seen, the function found failed the test; it is not a valid stream function for this field.

Note also that **find_stream_f2D()** integrates over x first and then over y, which may not always be the optimal sequence. In some cases, it may be necessary to reverse the order to find the stream function. This can be achieved by simply swapping the components V_x and V_y without altering the code.

4.6.7 *Case study: Constructing divergence-free fields*

Assume the component functions of 2D fields have the following characteristics:

- **Case 1:** One of the components is constant.
- **Case 2:** Both component functions are linear.

Our goal is to find the possible formulas for the divergence-free fields and their corresponding stream functions for these two cases.

Solution: In Case 1, let's assume the x-component of the vector field is constant, denoting it as a symbolic variable c, while the y-component is an unknown function defined by a symbolic function F_y. We can then utilize **find_stream_f2D()** to determine the stream function. Following is the code snippet to perform this task:

```
1  x, y, c = symbols('x, y, c', real=True)          # define variables
2  Fy = sp.Function('F_y')(y)                    # define the nuknown function
3  Fx = sp.Function('F_x')(x)
4  Vx = c; Vy = Fy
5  ψ = find_stream_f2D(Vx, Vy, X)
6  display(Math(f" \\text{{The stream function ψ}}={latex(ψ)}"))
```

ψ = c*y
(ψ.diff(x)+Vy).simplify() = F_y(y)
(ψ.diff(y)-Vx).simplify() = 0

The stream function $\psi = cy$

It is evident that the function found serves as a stream function for the field when $F_y = 0$. Therefore, the stream function can be expressed as

$$\psi(x, y) = cy \quad \text{or} \quad \psi(x, y) = -cx \tag{4.54}$$

where c is an arbitrary constant. The negative sign appears in the second equation due to the relationship defined in Eq. (4.51). The corresponding vector field functions are given by

$$\mathbf{F}(x, y) = \begin{bmatrix} c \\ 0 \end{bmatrix} \quad \text{or} \quad \mathbf{F}(x, y) = \begin{bmatrix} 0 \\ c \end{bmatrix} \tag{4.55}$$

Let us confirm all these results using the following code snippets:

```
1  Vx = c; Vy = 0
2  ψ = find_stream_f2D(Vx, Vy, X)
3  display(Math(f" \\text{{The stream function ψ}}={latex(ψ)}"))
```

ψ = c*y
(ψ.diff(x)+Vy).simplify() = 0
(ψ.diff(y)-Vx).simplify() = 0

The stream function $\psi = cy$

```
1  Vx = 0; Vy = c
2  ψ = find_stream_f2D(Vx, Vy, X)
3  display(Math(f" \\text{{The stream function ψ}}={latex(ψ)}"))
```

ψ = -c*x
(ψ.diff(x)+Vy).simplify() = 0
(ψ.diff(y)-Vx).simplify() = 0

The stream function $\psi = -cx$

We draw the following conclusion:

> If any one of the component functions of a vector field is constant and the other component is zero, the field qualifies as a divergence-free field. In this case, the stream function can be expressed as either $-cx$ or cy.

Solution: Now, let us consider Case 2. We will assume that the two components of the vector field are independent general linear functions. We can define these functions symbolically and subsequently use the find_stream_f2D() function to determine the stream functions. The code snippet for this is provided as follows:

```
 1  x, y = symbols('x, y', real=True)              # define variables
 2  a1, b1, c1 = symbols('a1, b1, c1', real=True)  # define constants
 3  a2, b2, c2 = symbols('a2, b2, c2', real=True)
 4  Fx = sp.Function('F_x')(x,y)                    # define the nuknown functions
 5  Fy = sp.Function('F_y')(x,y)
 6
 7  Fx = a1*x + b1*y + c1
 8  Fy = a2*x + b2*y + c2
 9  ψ = find_stream_f2D(Fx, Fy, X)
10  display(Math(f" \\text{{The stream function ψ}}={latex(ψ)}"))
```

ψ = a1*x*y - a2*x**2/2 + b1*y**2/2 + c1*y - c2*x
(ψ.diff(x)+Vy).simplify() = y*(a1 + b2)
(ψ.diff(y)-Vx).simplify() = 0

The stream function $\psi = a_1 xy - \dfrac{a_2 x^2}{2} + \dfrac{b_1 y^2}{2} + c_1 y - c_2 x$

The results indicate that if $a_1 = -b_2$, the derived function qualifies as a stream function. Due to the symmetry property, it follows that if $b_1 = -a_2$, the corresponding function will also serve as a stream function. Therefore, we obtain the following expressions for the stream functions:

$$\psi(x, y) = -\frac{a_2 x^2}{2} + \frac{b_1 y^2}{2} + c_1 y + x\,(a_1 y - c_2)$$

or   (4.56)

$$\psi(x, y) = -\frac{a_1 x^2}{2} + \frac{a_1 y^2}{2} + c_2 y - x\,(b_1 y + c_1)$$

The corresponding vector field functions are given by

$$\mathbf{F}(x, y) = \begin{bmatrix} a_1 x + b_1 y + c_1 \\ a_2 x - a_1 y + c_2 \end{bmatrix} \quad \text{or} \quad \mathbf{F}(x, y) = \begin{bmatrix} -b_1 x + b_2 y + c_2 \\ a_1 x + b_1 y + c_1 \end{bmatrix} \quad (4.57)$$

where a_1, b_1, c_1 and a_2, b_2, c_2 are arbitrary constants. We will verify these findings using the following code snippets:

```
 1  Fx = a1*x + b1*y + c1
 2  Fy = a2*x - a1*y + c2
 3  ψ = find_stream_f2D(Fx, Fy, X)
 4  display(Math(f" \\text{{The stream function ψ}}={latex(ψ)}"))
```

ψ = -a2*x**2/2 + b1*y**2/2 + c1*y + x*(a1*y - c2)
(ψ.diff(x)+Vy).simplify() = 0
(ψ.diff(y)-Vx).simplify() = 0

The stream function $\psi = -\dfrac{a_2 x^2}{2} + \dfrac{b_1 y^2}{2} + c_1 y + x\,(a_1 y - c_2)$

```
1  Fx =-b1*x + b2*y + c2              # swap Fx, and Fy, and then set b1=-a2
2  Fy = a1*x + b1*y + c1
3  ψ = find_stream_f2D(Fx, Fy, X)
4  display(Math(f" \\text{{The stream function ψ}}={latex(ψ)}"))
```

```
ψ = -a1*x**2/2 + b2*y**2/2 + c2*y - x*(b1*y + c1)
(ψ.diff(x)+Vy).simplify() = 0
(ψ.diff(y)-Vx).simplify() = 0
```

The stream function $\psi = -\dfrac{a_1 x^2}{2} + \dfrac{b_2 y^2}{2} + c_2 y - x\left(b_1 y + c_1\right)$

We draw the following conclusion:

> If the component functions are linear, the vector field is a divergence-free vector field if the coefficient of x in first component function is equal to the negative coefficient of y in the other component function.

Readers may compare this case study with that in Section 3.4.11 of Chapter 3.

4.6.8 *A list of often used divergence-free fields*

The following is a list of often used divergence-free fields provided by ChatGPT:

Field	Context
Magnetic field $\mathbf{B}$	Absence of magnetic monopoles
Incompressible fluid flow velocity field $\mathbf{v}$	In fluid dynamics
Electric field in steady-state current $\mathbf{E}$	Without free charge, steady-state
Vorticity field $\boldsymbol{\omega}$	In fluid dynamics
Magnetic induction field $\mathbf{H}$	In magnetostatics
Current density field $\mathbf{J}$	In steady-state conditions
Vector potential field $\mathbf{A}$	In electromagnetism
Magnetic flux density field in superconductors	In type-II superconductors
Momentum density field $\mathbf{p}$	In incompressible steady flows
Magnetization field $\mathbf{M}$	In ferromagnetic materials

$$(4.58)$$

4.7 Concluding remarks

This chapter studies Green's theorem for two-dimensional vector fields and its applications in detail. A number of formulas, techniques, and Python codes have been presented for examining Green's theorem. We close the chapter with the following remarks:

1. There are two versions of Green's theorem. Green's circulation theorem captures the circulation of a vector field in a domain, which equals the work done by the field along the boundary of the domain. Green's divergence theorem, on the other hand, captures the divergence of a vector field in a domain, which equals the flux of the field across the boundary of the domain. Mathematically, both are identical.
2. Green's theorem is a mathematical statement of conservation laws. For example, the circulation theorem represents an energy conservation law, while the divergence theorem represents a mass conservation law.
3. Python functions are provided for examining both versions of Green's theorem for a given vector field, using circular domains and their boundaries.
4. Using Green's theorem, the area of a 2D domain can be computed using a curve integral. Formulas and algorithms are presented for computing the area of polygons with an arbitrary number of edges.
5. Conditions for divergence-free vector fields are discussed, including a Python function to find the stream function for a given divergence-free vector field, which complements the potential function discussed in the previous chapter.

This chapter is dense, but it provides a solid foundation for the subsequent chapters. The following chapter will discuss surface integrals of vector functions defined in three-dimensional space.

There are special vector fields that are both conservative and divergence-free, which will be discussed in detail in Chapter 9.

References

[1]   G. Strang, *Calculus*, Wellesley-Cambridge Press, Massachusetts, 1991.
[2]   G.R. Liu, *Numbers and Functions: Theory, Formulation, and Python Codes*, World Scientific, New Jersey, 2024.
[3]   G.R. Liu, *Mechanics of Materials: Formulations and Solutions with Python*, World Scientific, New Jersey, 2024.
[4]   G.R. Liu, *Calculus: A Practical Course with Python*, World Scientific, New Jersey, 2025.

Chapter 5

Surface Integrals

```
 1  # Place cursor in this cell, and press Ctrl+Enter to import dependences.
 2  import sys                        # For accessing the computer system
 3  sys.path.append('../grbin/')          # Add in the path to your system
 4
 5  from commonImports import *       # Import dependences from '../grbin/'
 6  import grcodes as gr                # Import the module of the author
 7  importlib.reload(gr)             # When grcodes is modified, reload it
 8
 9  init_printing(use_unicode=True)       # For latex-like quality printing
10  np.set_printoptions(precision=4,suppress=True,
11                      formatter={'float_kind': '{:.4e}'.format})
```

In the previous chapters, Green's theorem for two-dimensional vector fields was discussed in detail with applications involving curve integrals and 2D domain integrals. This chapter studies surface integrals of various types of 3D scalar and vector functions in detail with applications. This complements the surface integrals for scalar functions discussed in Chapter 5 of Ref. [1]. Our focus in this chapter will be on techniques for computing the normal vectors of 3D surfaces as well as the fluxes of vector fields across 3D surfaces.

This chapter is written with reference to textbooks in Refs. [2–4]. Wikipedia pages, particularly those on integral (https://en.wikipedia.org/wiki/Integral), surface integral (https://en.wikipedia.org/wiki/Surface_integral), and Green's theorem (https://en.wikipedia.org/wiki/Green%27s_theorem), serve as valuable additional references.

Both NumPy and SymPy are used in the development of code for the demonstration examples. Discussions with ChatGPT, Gemini, and Bing have also greatly helped in coding and in the preparation of this chapter.

5.1 Surface integral of vector functions

If the integrand is a vector function, the surface integral must account for its directional effects. Figure 5.1 shows a schematic drawing of a vector field on a surface $\mathcal{S}$, where we consider the dot product of the vector field with the normal vector of the surface to measure these effects.

5.1.1 *Formulation*

Consider a vector field denoted as $\mathbf{V}$ on a smooth (differentiable) surface $\mathcal{S}$. A position vector $\mathbf{r} = (x, y, z)$ is used to define any point on the surface $\mathcal{S}$, as shown in Fig. 5.2.

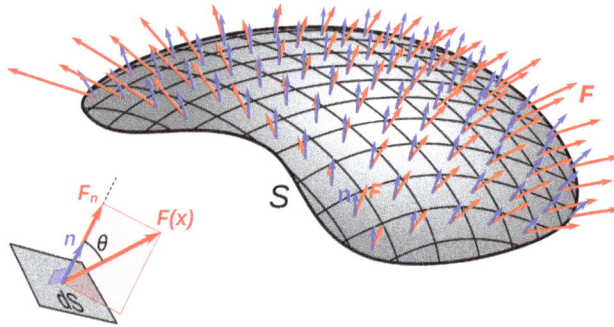

Figure 5.1. Schematic illustration of a vector field $\mathbf{F}$ on surface $\mathcal{S}$.

Source: Image from Wikimedia Commons by Chetvorno under the CC0 license.

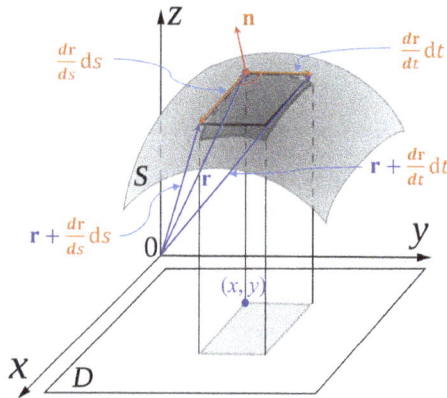

Figure 5.2. Schematic illustration of a differential surface parallelogram and position vectors defining a point on surface $\mathcal{S}$.

Source: Image adapted from Wikimedia Commons by Cronholm144 under the CC BY-SA 3.0 license.

The position vector can be written generally as

$$r(x, y, z) = \mathbf{r}\big(x(s,t), y(s,t), z(s,t)\big) = \mathbf{r}(s,t) \tag{5.1}$$

In this case, the given vector function can be expressed as a function of the position vector: $\mathbf{V}(\mathbf{r})$. Assuming $\mathbf{V}$ represents a velocity field, the flux across the surface S, denoted as q, can be defined as

$$q = \iint_S \mathbf{V} \cdot \mathbf{n} dS = \iint_S (\mathbf{V} \cdot \mathbf{n}) dS$$

$$= \iint_{D_p} \left(\mathbf{V}(\mathbf{r}(s,t)) \cdot \frac{\frac{\partial \mathbf{r}}{\partial s} \times \frac{\partial \mathbf{r}}{\partial t}}{\left\| \frac{\partial \mathbf{r}}{\partial s} \times \frac{\partial \mathbf{r}}{\partial t} \right\|} \right) \left\| \frac{\partial \mathbf{r}}{\partial s} \times \frac{\partial \mathbf{r}}{\partial t} \right\| ds\, dt \tag{5.2}$$

$$= \iint_{D_p} \mathbf{V}(s,t) \cdot \left(\frac{\partial \mathbf{r}}{\partial s} \times \frac{\partial \mathbf{r}}{\partial t} \right) dsdt = \iint_{D_p} \mathbf{V}(s,t) \cdot \mathbf{n} dsdt$$

This is a one-dimensional extension of the 2D counterpart discussed in Section 3.6.

The domain D_p is defined with (s,t) coordinates corresponding to the surface S. To determine the differential area on the surface, we use the cross-product of the two vectors $\frac{d\mathbf{r}}{ds}ds$ and $\frac{d\mathbf{r}}{dt}dt$. This cross-product gives the area of the differential parallelogram, denoted as dS, as shown in Fig. 5.2. Here, $\left| \frac{\partial \mathbf{r}}{\partial s} \times \frac{\partial \mathbf{r}}{\partial t} \right|$ represents the scalar area of this parallelogram.

Using Eq. (5.1), we have

$$\frac{\partial \mathbf{r}}{\partial s} = \begin{bmatrix} \frac{\partial x}{\partial s} \\ \frac{\partial y}{\partial s} \\ \frac{\partial z}{\partial s} \end{bmatrix} ; \quad \frac{\partial \mathbf{r}}{\partial t} = \begin{bmatrix} \frac{\partial x}{\partial t} \\ \frac{\partial y}{\partial t} \\ \frac{\partial z}{\partial t} \end{bmatrix} \tag{5.3}$$

The normal vector $\mathbf{n}$ is given by the cross-product [4]:

$$\mathbf{n} = \frac{\partial \mathbf{r}}{\partial s} \times \frac{\partial \mathbf{r}}{\partial t} = \begin{vmatrix} \mathbf{i} & \mathbf{j} & \mathbf{k} \\ \frac{\partial x}{\partial s} & \frac{\partial y}{\partial s} & \frac{\partial z}{\partial s} \\ \frac{\partial x}{\partial t} & \frac{\partial y}{\partial t} & \frac{\partial z}{\partial t} \end{vmatrix} = \begin{bmatrix} \frac{\partial y}{\partial s}\frac{\partial z}{\partial t} - \frac{\partial z}{\partial s}\frac{\partial y}{\partial t} \\ \frac{\partial z}{\partial s}\frac{\partial x}{\partial t} - \frac{\partial x}{\partial s}\frac{\partial z}{\partial t} \\ \frac{\partial x}{\partial s}\frac{\partial y}{\partial t} - \frac{\partial y}{\partial s}\frac{\partial x}{\partial t} \end{bmatrix} \tag{5.4}$$

The detailed derivation for the expression for $\mathbf{n}$ through a cross-product can be found in Ref. [1].

Equation (5.2) produces a scalar. This implies the following:

1. If $\mathbf{V}$ is tangent to S at a point, no flux will flow across the surface S at that point, resulting in zero flux there.

2. If $\mathbf{V}$ is perpendicular to S at a point, the flux will be $|\mathbf{n}|dS$. It will be positive if $\mathbf{V}$ is in the same direction as $\mathbf{n}$; otherwise, it will be negative.
3. In any case, the ins and outs of the flux at any location on the surface are automatically measured by the dot product. The integral sums all the flux over the entire surface, yielding a total flux.

Note that in Eq. (5.2), the norm computation is canceled, eliminating the difficulty related to the square roots from norm calculations, which typically occur in line or surface integrals, as seen multiple times in Ref. [1]:

> It is clear that a surface integral of a vector function involves the interaction between two directions: the direction of the vector function and that of the surface. This interaction is their dot product, resulting in a scalar. This is the major difference between the surface integral of a scalar function and that of a vector function.

When the surface is given as $z = f(x, y)$ and the vector field is given as $\mathbf{V} = [V_x \quad V_y \quad V_z]^\top$, the flux can be computed using a simpler formula:

$$q = \iint_S \mathbf{V} \cdot \mathbf{n}dS = \iint_{D_p} \mathbf{V}(x, y, z) \cdot \mathbf{n}dxdy$$

$$= \iint_{D_p} \left(-V_x \frac{\partial z}{\partial x} - V_y \frac{\partial z}{\partial y} + V_z \right) dxdy \tag{5.5}$$

in which we used $\mathbf{n}$ that is given as [1]

$$\mathbf{n} = \frac{\partial \mathbf{r}}{\partial x} \times \frac{\partial \mathbf{r}}{\partial y} = \begin{vmatrix} \mathbf{i} & \mathbf{j} & \mathbf{k} \\ 1 & 0 & \frac{\partial z}{\partial x} \\ 0 & 1 & \frac{\partial z}{\partial y} \end{vmatrix} = \begin{bmatrix} -\frac{\partial z}{\partial x} \\ -\frac{\partial z}{\partial y} \\ 1 \end{bmatrix} \tag{5.6}$$

Let us look at some examples.

5.1.2 *Example: Flux of a constant vector field over a surface*

Consider first a simple constant vector field given as

$$\mathbf{V} = \begin{bmatrix} c_x & c_y & c_z \end{bmatrix}^\top \tag{5.7}$$

where c_x, c_y, and c_z are given constants. Assume the surface S is spherical:

$$x^2 + y^2 + z^2 = a^2 \tag{5.8}$$

where a is the radius of the sphere.

1. Compute the flux of the field through the upper and lower halves of the spherical surface.
2. Compute the flux of the field through the entire spherical surface.

Solution: Since the surface is a sphere, we should use spherical coordinates for this computation. The mapping between spherical and Cartesian coordinates is discussed in detail in Ref. [1]. The unit normal vector **n** can then be computed using Eq. (5.4) with θ and ϕ as the spherical coordinates. These formulas are directly used in the following code:

```
1  # Define variables and vector function:
2  x, y, z, θ, φ = symbols('x, y, z, θ, φ', real=True)
3  cx, cy, cz = symbols('c_x, c_y, c_z ', real=True)
4  a = symbols('a', positive=True)
5
6  Vc = Matrix([cx, cy, cz])                    # the constant vector field
```

To compute the integral, we write the following Python function for repeated use:

```
1  def flux_sphere(V, a, x, y, z, θ, φ):
2      '''
3      Compute the flux of a given vector V across a spherical surface
4      with radius a.
5          inputs: V - sympy, the given vector function
6                  x, y, z - symbolic variables for Cartesian coordinates
7                  θ, φ - symbolic variables for spherical coordinates
8                  a - radius of the sphere
9          return: n - normal vector on S
10                  I_f_U - Flux of V through upper spherical surface
11                  I_f_L - Flux of V through lower spherical surface
12                  I_f_F - Flux of V through entire spherical surface
13      '''
14      # Sphere - Cartesian coordinates mapping:
15      x_ = a*sin(θ)*cos(φ)    # x_, y_, z_ are x,y,z on the sphere surface
16      y_ = a*sin(θ)*sin(φ)
17      z_ = a*cos(θ)
18
19      r_θ = Matrix([x_.diff(θ), y_.diff(θ), z_.diff(θ)])      # Jacobian
20      r_φ = Matrix([x_.diff(φ), y_.diff(φ), z_.diff(φ)])
21      n = gr.cross2vectors(r_θ, r_φ, sympy=True)      # normal vector on S
22
23      dic = {x:x_, y:y_, z:z_}; pi = sp.pi
24      I_f_U = integrate(integrate(V.subs(dic).T@n, (φ,0,2*pi)),(θ,0,pi/2))
25      display(Math(f" \\text{{Flux of V through upper spherical surface\
26                                      = }}{{latex(I_f_U[0])}}"))
27
```

```
28      I_f_L = integrate(integrate(V.subs(dic).T@n,(φ,0,2*pi)),(θ,pi/2,pi))
29      display(Math(f" \\text{{Flux of V through lower spherical surface\
30                                    = }}{{latex(I_f_L[0])}}"))
31
32      I_f_F = integrate(integrate(V.subs(dic).T@n, (φ,0,2*pi)),(θ,0,pi))
33      display(Math(f" \\text{{Flux of V through entire spherical surface\
34                                    = }}{{latex(I_f_F[0])}}"))
35      return n, I_f_U, I_f_L, I_f_F
```

Now, we use flux_sphere() to find the solutions:

```
1  n_sphere, _, _, _ = flux_sphere(Vc, a, x, y, z, θ, φ)
```

Flux of V through upper spherical surface $= \pi a^2 c_z$

Flux of V through lower spherical surface $= -\pi a^2 c_z$

Flux of V through entire spherical surface $= 0$

We found the following:

1. The flux of the constant vector field through the upper half of the spherical surface is the strength of the field in the vertical z-direction, c_z, multiplied by the area of the half-spherical surface projected along the z-direction onto the (x, y) plane.
2. The flux of the constant vector field through the lower half of the spherical surface is the strength of the field in the negative vertical z-direction, $-c_z$, multiplied by the area of the half-spherical surface projected along the z-direction onto the (x, y) plane.
3. The flux of the constant vector field through the entire spherical surface is zero.

This is because the outward normal vector on the spherical surface changes direction, while the constant vector field remains the same. The dot product of these two vectors accounts for the directional difference. Alternatively, the constant vector field is antisymmetric with respect to the x–y plane, and the entire spherical surface (including its orientation) is symmetric, leading to a zero integral value due to the symmetry property discussed in Chapter 4 of Ref. [1].

Alternative (or better) explanations can be made in the following chapter when discussing the divergence theorem for 3D problems.

Readers are encouraged to explore the differences in flux through the left and right halves of the spherical surface.

5.1.3 *Example: Flux of the position vector field through a sphere*

Consider the position vector field given as

$$\mathbf{V} = \begin{bmatrix} x & y & z \end{bmatrix}^{\top} \qquad (5.9)$$

All other settings remain unchanged from the previous example. Let's conduct the same study as before.

Solution: Analysis of the problem. Since the only change is the vector field, many lines of code can be reused directly. Additionally, this position vector field is point symmetric with respect to the origin of the coordinates. Given that the spherical surface is also point symmetric, we can anticipate the following:

1. The integral will always yield a positive value as long as the integrated surface area is not zero.
2. The flux should be proportional to the area of the surface being integrated.

This analysis provides a partial solution. Let us write the following code snippet to compute the precise results:

```
1  Vp = Matrix([x, y, z])                    # the position vector field
2
3  n_sphere, _, _, _ = flux_sphere(Vp, a, x, y, z, θ, φ)
```

Flux of V through upper spherical surface $= 2\pi a^3$

Flux of V through lower spherical surface $= 2\pi a^3$

Flux of V through entire spherical surface $= 4\pi a^3$

From these results, we observed the following:

1. Regardless of whether the upper or lower spherical surface is integrated, we obtain the same result as expected.
2. When the entire spherical surface is used, the integral value is doubled, which is also anticipated.

The position vector field radiates outward in all directions similar to a point light source. The flux through a part of a spherical surface should be proportional to the area of that partial surface. This demonstrates that both the direction of the vector field and the direction and area of the surface are crucial to determining the integral's value.

The example in the following section further illustrates this point.

5.1.4 *Example: Compute the flux out of a cone*

Consider the same position vector field given in Eq. (5.9) as the vector velocity field.

This time, we consider a **cone surface** as the integral surface $\mathcal{S}$. It is defined as

$$z = \sqrt{x^2 + y^2} \tag{5.10}$$

Compute the flux of the velocity field through the cone surface.

Solution: In this case, the surface is defined in terms of z. We will use Eq. (5.6) to compute the normal vector. All other computations are similar to those in the previous case. We write the following code to determine the solution:

```
1  x, y, z = symbols('x, y, z', real=True)
2
3  Vp = Matrix([x, y, z])                           # the vector field
4  z_ = sp.sqrt(x**2 + y**2)      # x_, y_, and z_ are x,y,z on the surface
5  n = Matrix([-z_.diff(x), -z_.diff(y), 1])
6
7  dic = {z:z_}
8  I_f = integrate(integrate(Vp.subs(dic).T@n, (x, 0, 1)), (y, 0, 1))
9  print(f" Flux out of the cone surface = {I_f[0]}")
10 display(Math(f" \\text{{The dot-product is: }}\
11     {latex((Vp.subs(dic).T@n)[0].simplify())}"))
```

```
Flux out of the cone surface = 0
```

The dot product: 0

We found that the flux through the cone surface is zero. This is because the velocity field is perpendicular to the normal vector of the cone surface. The field flows along the cone surface, so no flux crosses the cone surface. This demonstrates again that the direction of the field relative to the surface normal is crucial.

To emphasize this point further, let us slightly change the direction of the vector field and compute the flux:

```
1  Vp2 = Matrix([x, y, z/2])          # the direction changed vector field
2
3  I_f = integrate(integrate(Vp2.subs(dic).T@n, (x, 0, 1)), (y, 0, 1))
4  print(f" Flux out of the cone surface = {I_f[0].evalf(5)}")
5  display(Math(f" \\text{{The dot-product is: }}\
6     {latex((Vp2.subs(dic).T@n)[0].simplify())}"))
```

Flux out of the cone surface = -0.38260

The dot product is $-\dfrac{\sqrt{x^2 + y^2}}{2}$

We found that the dot product is no longer zero and the surface integral yields a nonzero value.

Now, let's consider an example using cylindrical coordinates.

5.1.5 *Example: Flux of velocity field through a cylinder*

Consider the same position vector given in Eq. (5.9) as the velocity field.

The integral surface is a cylinder defined as

$$x^2 + y^2 = a^2 \tag{5.11}$$

where a is the radius of the cylinder.

Compute the flux of the velocity field through the cylinder in the region $0 \le z \le h$, where h is the height.

Solution: Since the integral surface is now a cylinder, we shall use cylindrical coordinates [1] for this task. We write the following code to compute the flux through the cylinder:

```
1  x, y, θ, t = symbols('x, y, θ, t', real=True)
2  a, h = symbols('a, h', positive=True)
3
4  Vp = Matrix([x, y, z])                              # the vector integrand
5
6  # Cylindric - Cartesian coordinate mapping:
7  x_ = a*cos(θ)          # x_, y_, and z_ are x,y,z on the cylinder surface
8  y_ = a*sin(θ)
9  z_ = t
10
11 r_θ = Matrix([x_.diff(θ), y_.diff(θ), z_.diff(θ)])          # Jacobian
12 r_t = Matrix([x_.diff(t), y_.diff(t), z_.diff(t)])
13 n = gr.cross2vectors(r_θ, r_t, sympy=True)         # normal vector on S
14
15 dic = {x:x_, y:y_, z:z_}
16 I_f = integrate(integrate(Vp.subs(dic).T@n, (θ,0,2*sp.pi)), (t, 0, h))
17
18 display(Math(f" \\text{{Flux through a cylinder = }}{latex(I_f[0])}"))
```

Flux through a cylinder $= 2\pi a^2 h$

The result shows that the flux through the lateral surface of the cylinder is the field strength a (where $a = \sqrt{x^2 + y^2}$) multiplied by the lateral surface area, which is $2\pi a h$.

5.1.6 *On integrals over nondifferentiable surfaces*

When computing the flux through the top surface of the cylinder, we cannot use the same formula for **n** as used for the lateral surface because the top surface is flat and defined by $z = h$. The complete surface of the cylinder, including the top, is not differentiable along the circular edge (which is a curve with kink points). Therefore, we must treat the top surface as a separate piece and perform piecewise integration.

For the top surface of the cylinder where $z = h$, the unit normal vector is $\mathbf{n} = [0, 0, 1]^\top$. Therefore, $\mathbf{V} \cdot \mathbf{n} = h$. The flux through the top surface is h multiplied by the area of the top circle, which is πa^2, giving $\pi a^2 h$. For the bottom surface where $z = 0$, the flux is zero: $\pi a^2 \times 0 = 0$.

Thus, the total flux across the combined surface of the cylinder, including both the top and bottom caps, is $2\pi a^2 h + \pi a^2 h + 0 = 3\pi a^2 h$.

5.1.7 *Example: Flux of a spin field out of a sphere*

Consider the velocity field defined as

$$\mathbf{s}_x = \begin{bmatrix} 0 & -z & y \end{bmatrix}^\top \tag{5.12}$$

The integral surface is still the spherical surface defined earlier.

Compute the flux of the velocity field through the top surface of the sphere.

Solution: We write the following code to get this done:

```
1  Vsx = Matrix([0, -z, y])                          # the spin vector field
2
3  n_sphere, _, _, _ = flux_sphere(Vsx, a, x, y, z, θ, φ)
```

Flux of V through upper spherical surface $= 0$

Flux of V through lower spherical surface $= 0$

Flux of V through entire spherical surface $= 0$

We found that the flux is zero for all these cases computed. This is because the spin field rotates parallel to the surface of any spherical surface. There is naturally no flux across the surface.

If the velocity field is a mixture of the position and spin fields, e.g.,

$$\mathbf{s}_x = \begin{bmatrix} x & -z & y \end{bmatrix}^\top \tag{5.13}$$

The first component x is a component of the position vector field, while the remaining two components correspond to the spin field. The flux should pick up the contribution from the position vector field. The following code confirms this prediction:

```
1  Vps = Matrix([x, -z, y])                          # the spin vector field
2
3  n_sphere, _, _, _ = flux_sphere(Vps, a, x, y, z, θ, φ)
```

Flux of V through upper spherical surface $= \dfrac{2\pi a^3}{3}$

Flux of V through lower spherical surface $= \dfrac{2\pi a^3}{3}$

Flux of V through entire spherical surface $= \dfrac{4\pi a^3}{3}$

The fluxes are one-third of the flux of the full position vector field obtained in the previous example. None of the spin components are captured.

We will see more examples in the following chapter when discussing the divergence of a vector field, as the divergence inside a domain is directly related to the flux across the boundary surface of the domain similar to the discussions given in the previous chapter.

5.2 Concluding remarks

This chapter studies in detail surface integrals of various types of 3D scalar and vector functions with applications. A number of formulas, techniques, and Python codes have been presented for computing integrals over surfaces. The integrals produce important geometric features of 3D surfaces and the fluxes of vector fields across 3D surfaces. We have developed formulations and Python codes for the following tasks:

1. fluxes of vector fields across 3D surfaces, including circular, spherical, and cylindrical surfaces;
2. a large number of examples have been presented in addition to examples on the integral of scalar functions over continuous surfaces given in Ref. [1].

The following chapter discusses how a surface integral is related to a 3D domain integral for vector fields: the divergence theorem, which is an extension of the 2D Green's divergence theorem.

References

[1] G.R. Liu, *Calculus: A Practical Course with Python*, World Scientific, New Jersey, 2025.
[2] G. Strang, *Calculus*, Wellesley-Cambridge Press, Massachusetts, 1991.
[3] G.R. Liu, *Numbers and Functions: Theory, Formulation, and Python Codes*, World Scientific, New Jersey, 2024.
[4] G.R. Liu, *Mechanics of Materials: Formulations and Solutions with Python*, World Scientific, New Jersey, 2024.

Chapter 6

The Divergence Theorem

```
 1  # Place cursor in this cell, and press Ctrl+Enter to import dependences.
 2  import sys                        # For accessing the computer system
 3  sys.path.append('../grbin/')          # Add in the path to your system
 4
 5  from commonImports import *        # Import dependences from '../grbin/'
 6  import grcodes as gr                  # Import the module of the author
 7  importlib.reload(gr)              # When grcodes is modified, reload it
 8
 9  init_printing(use_unicode=True)       # For latex-like quality printing
10  np.set_printoptions(precision=4,suppress=True,
11                      formatter={'float_kind': '{:.4e}'.format})
```

In the previous chapter, the surface integrals of 3D scalar and vector functions were discussed along with techniques and codes for computing the fluxes of vector fields across 3D surfaces. This chapter explores how surface integrals are related to 3D domain integrals for vector fields, leading to the divergence theorem, which is an extension of the 2D Green's theorem discussed in Chapter 4. Techniques and codes will be presented for examining the divergence theorem for various types of vector fields.

This chapter is written with reference to the textbooks in Refs. [1–3]. Wikipedia pages, particularly those on surface integral (https://en.wikipedia.org/ wiki/Green%27s_theorem) and divergence theorem (https://en.wikipedia. org/wiki/Divergence_theorem), serve as valuable additional references.

Both NumPy and SymPy are used in the development of code for the demonstration examples. Discussions with ChatGPT, Gemini, and Bing have also greatly helped in coding and in the preparation of this chapter.

149

6.1 The expression

Consider a differentiable vector field, defined in a 3D domain $\mathcal{D}$ bounded by a smooth (or piecewise smooth with a finite number of pieces) surface boundary $\mathcal{S}$, as shown in Fig. 6.1.

The expression for the vector field is given as

$$\mathbf{V} = \begin{bmatrix} V_x(x,y,z) \\ V_y(x,y,z) \\ V_z(x,y,z) \end{bmatrix} = \begin{bmatrix} M(x,y,z) \\ N(x,y,z) \\ P(x,y,z) \end{bmatrix} \tag{6.1}$$

where V_x, V_y, and V_z are the three component functions. These are often denoted in the literature as M, N, and P, respectively. Across the surface boundary $\mathcal{S}$, the medium can flow in and out. The divergence theorem relates the domain integral of a vector field over a 3D domain to the surface integral of the field across the boundary of the domain. It is expressed as

$$\oint_{\mathcal{S}} (\mathbf{V} \cdot \mathbf{n}) \mathrm{d}S = \underbrace{\iiint_{\mathcal{D}} \left(\frac{\partial V_x}{\partial x} + \frac{\partial V_y}{\partial y} + \frac{\partial V_z}{\partial z} \right)}_{\mathrm{div}\mathbf{V}} \mathrm{d}x\,\mathrm{d}y\,\mathrm{d}z \tag{6.2}$$

or simply

$$\underbrace{\oint_{\mathcal{S}} \mathbf{V} \cdot \mathbf{n} \mathrm{d}S}_{\text{total flux}} = \underbrace{\iiint_{\mathcal{D}} \underbrace{\nabla \cdot \mathbf{V}}_{\mathrm{div}\mathbf{V}} \mathrm{d}V}_{\text{total divergence}} \tag{6.3}$$

Here, we require all the integrands being integrable for the equality to hold. We also require that the domain is simply connected, as discussed in Section 4.1.1 of Chapter 4. If the domain is multiply connected, cuts are needed to make it simply connected.

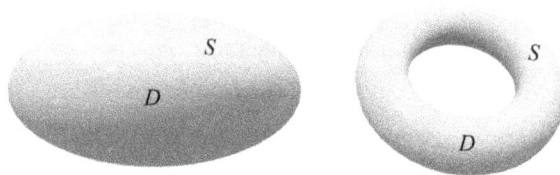

Figure 6.1. A schematic drawing of a three-dimensional (3D) body occupying domain $\mathcal{D}$ and bounded by its surface $\mathcal{S}$. Left: a simply connected domain. Right: a typical multiply-connected domain, which requires cuts to make it simply connected.

Before explaining the terms in the preceding equation, let us have a simple analysis:

1. If $\mathbf{V} = 0$, we have $0 = 0$. The equation obviously holds.
2. If $\mathbf{V} = \mathbf{c}$, where $\mathbf{c}$ is a constant vector, the equation holds for the following reasons. On the right side, $\nabla \cdot \mathbf{V} = 0$ due to the differentiation. On the left side, $\oint_S \mathbf{c} \cdot \mathbf{n} dS$ is zero because the boundary integral is closed, and the result is the sum of all normal vectors. When integrating around the entire smooth surface, it vanishes. This represents a simple **divergence-free** field, which is discussed in detail in the previous chapter.

Equation (6.3) is an extension of the 2D version, Eq. (4.37). Let us examine the meanings of the terms in Eq. (6.3).

6.1.1 *A statement of a conservation law in 3D*

The left side represents the closed surface integral over the surface boundary S of domain $\mathcal{D}$ and $\mathbf{n}$ is the outward normal vector on S. Often, $\mathbf{n} dS$ is written as $d\mathbf{S}$. The physical meaning of this term depends on the type of physical problem. For example, if the vector field is a velocity field, $\oint_S \mathbf{V} \cdot \mathbf{n} dS$ represents the flux (the amount of flow) of the vector field across the surface S that encloses domain $\mathcal{D}$. Due to the dot product, the flow along the boundary surface (circulating along the boundary) is not taken into account. Therefore, it represents the net flow in and out across the surface and is a scalar value, as studied in great detail in Chapter 4.

The right side is a volume integral over the domain $\mathcal{D}$. It measures the sum of all divergences in $\mathcal{D}$. If the vector field is a velocity field, it is a measure of the distributed sources in $\mathcal{D}$. Therefore, this must equal the flux across the closed boundary S. In essence, the divergence theorem is a mathematical description of the mass conservation law similar to the 2D case discussed in Section 4.5 of Chapter 4.

A rigorous proof of the conservation law for the 2D case is given in Section 4.5 of Chapter 4, which can be extended to 3D. Hence, we do not repeat it here.

6.1.2 *The first fundamental theorem of calculus in 3D*

Readers may compare the divergence theorem with the **first fundamental theorem of calculus** studied in Section 4.1.9 of Ref. [4] and its extension to 2D discussed in Section 4.4, in particular Fig. 4.6, of Chapter 4.

This comparison can provide a good understanding of how divergence in the 1D and 2D cases is canceled out with each other in the interior of the domain and why it equals the flux only on the boundary. The comparison may conclude that the divergence theorem is a dimensional extension of the first fundamental theorem of calculus as applied to 2D integration.

An intuitive explanation with a proof of the divergence theorem can be found in a Wikipedia page on divergence theorem (https://en.wikipedia.org/wiki/Divergence_theorem). We will not repeat these here. Instead, let us see some examples.

6.2 Case study: The divergence theorem over a sphere

In this case study, we aim to examine the following:

- the divergence of a given vector field **V** at any point in the domain,
- the total divergence of the vector field in a spherical domain with radius a,
- the volume integral of the divergence of the vector field over the sphere,
- the total flux of the vector field out of the sphere surface,
- check the satisfaction of the divergence theorem for various types of vector fields.

6.2.1 *Python code for the examination of vector fields*

For the convenience of our studies, we write the following Python function to examine a given vector field against the divergence theorem. The code is made self-explanatory with comments:

```
 1  def exam_DivThm_sphere(V, X, a):
 2      '''Examine a given vector field V that is differentiable over
 3          a spherical domain with radius a. The code prints out
 4          1 The divergence of the vector field in the domain.
 5          2 The total divergence of V over spherical domain with radius a.
 6          3 The total flux of the vector field across the sphere surface.
 7          4 Whether or not the divergence theorem holds.
 8      '''
 9      r = sp.symbols('r', nonnegative=True)  # radial coordinate, spheres
10      θ, φ = symbols('θ, φ', real=True)
11      x, y, z = X
12
```

```
13      #V_ball = 4*sp.pi*a**3/3 # volume of sphere of a; used for checking
14      div_V = (gr.div_vf(V, X)).simplify()     # divergence of the vector
15      display(Math(f" \\text{{Divergence of the vector field}}\
16                                        = {latex(div_V)}"))
17      # Integral over an enclosed sphere:
18      x_ = r*sin(θ)*cos(φ)     # x_, y_, and z_ are x,y,z on sphere surface
19      y_ = r*sin(θ)*sin(φ)
20      z_ = r*cos(θ)
21      dic = {x:x_, y:y_, z:z_}
22
23      # Volume integration over the spherical ball with r=a:
24      fID = div_V.subs(dic)*(r**2 *sin(θ))         # detJ = r**2 *sin(θ)
25      Int = sp.integrate; pi = sp.pi
26      div_VV = Int(Int(Int(fID,(r,0,a)),(θ,0,pi)),(φ,0,2*pi)).simplify()
27      display(Math(f"\\text{{ Integrand of the domain integral}}=\
28                                        {latex(sp.simplify(fID))}"))
29      # Surface integration over sphere of r=a:
30      n = X.subs(dic)/a          # normal n on S: n = 1/r on sphere surface
31      #display(Math(f" \\text{{ unit normal n = }}{latex(n)}"))
32
33      fSI = ((V.T@n)*(r**2 *sin(θ))).subs(dic).subs(r,a)[0].simplify()
34      #display(Math(f" \\text{{ Surface integrand = }}{latex(fSI)}"))
35      Vns=Int(Int(fSI,(φ,0,2*pi)),(θ,0,pi)).simplify()
36
37      display(Math(f" \\text{{Total flux of the vector field across the \
38                          sphere surface}} = {latex(Vns)}"))
39      display(Math(f" \\text{{Total divergence over the sphere}} \
40                                        = {latex(div_VV)}"))
41      print(f" Does the divergence theorem hold?  {Vns==div_VV}")
```

6.2.2 *Example: Divergence-free spin fields*

Consider spin fields

$$\mathbf{s}_x = \begin{bmatrix} 0 \\ -z \\ y \end{bmatrix}; \quad \mathbf{s}_y = \begin{bmatrix} z \\ 0 \\ -x \end{bmatrix}; \quad \mathbf{s}_z = \begin{bmatrix} -y \\ x \\ 0 \end{bmatrix}; \quad \mathbf{s}_z = \begin{bmatrix} -y + 2c \\ x + 3c \\ 4c \end{bmatrix} \quad (6.4)$$

The first three fields are 3D spin fields with respect to the x-, y-, and z-axes. The last field is also a spin field, but its center is shifted, with c being an arbitrary constant used to shift the center of the field.

Show the following:

1. These fields are divergence-free.
2. The surface integral over a sphere centered at the origin is zero, and hence the divergence theorem holds.

Solution: We use **exam_DivThm_sphere**() and the following code snippet to perform the task:

```
1  x, y, z = symbols('x, y, z', real=True)
2  a = symbols('a', nonnegative=True)              # radius of the sphere
3  X = Matrix([x, y, z])
4
5  sx = Matrix([0,-z, y])
6  # the given vector field
7  exam_DivThm_sphere(sx, X, a)
```

Divergence of the vector field $= 0$

Integrand of the domain integral $= 0$

Total flux of the vector field across the sphere surface $= 0$

Total divergence over the sphere $= 0$

Does the divergence theorem hold? True

```
1  # For sy:
2  sy = Matrix([z, 0, -x])                         # the given vector field
3  exam_DivThm_sphere(sy, X, a)
```

Divergence of the vector field $= 0$

Integrand of the domain integral $= 0$

Total flux of the vector field across the sphere surface $= 0$

Total divergence over the sphere $= 0$

Does the divergence theorem hold? True

```
1  # For sz:
2  sz = Matrix([-y, x, 0
3  exam_DivThm_sphere(sz, X, a)
```

Divergence of the vector field $= 0$

Integrand of the domain integral $= 0$

Total flux of the vector field across the sphere surface $= 0$

Total divergence over the sphere $= 0$

Does the divergence theorem hold? True

```
1  ## For sz with the spin center is shifted:
2  c = symbols('c', real=True)
3  sx = Matrix([-y+2*c, x+3*c, 4*c])
4  # the given vector field
5  exam_DivThm_sphere(sx, X, a)
```

Divergence of the vector field $= 0$

Integrand of the domain integral $= 0$

Total flux of the vector field across the sphere surface $= 0$

Total divergence over the sphere $= 0$

```
Does the divergence theorem hold? True
```

As seen, the divergence of the spin field is zero, and the flux of the spin field over any closed sphere is zero, and the divergence theorem holds. This is because the construction of the field results in zero partial derivatives for these component functions. The spin flow circulates with a net zero inflow and outflow, and this remains true regardless of whether the spin center is shifted. The divergence operator eliminates the effect of any constant term in the component functions.

We note the following:

> The spin fields are divergence-free regardless of where their center is located.

6.2.3 *Example: Position fields, constant divergences*

Consider position fields

$$
\mathbf{r}_1 = \begin{bmatrix} x \\ y \\ z \end{bmatrix}; \quad \mathbf{r}_2 = \begin{bmatrix} x \\ y \\ z + c/2 \end{bmatrix}; \quad \mathbf{r}_3 = \begin{bmatrix} x \\ y \\ z + 2c \end{bmatrix} \tag{6.5}
$$

where c is an arbitrary constant used for shifting the center of the field.

Show the following:

1. These fields have a constant divergence.
2. The surface integral over the sphere centered at the origin equals the volume integral, and the divergence theorem holds.

Solution: We use the following code to perform the task:

```
1  # Consider r1:
2
3  x, y, z = symbols('x, y, z', real=True)
4  a = symbols('a', positive=True)          # radius of the sphere
5
6  r1 = Matrix([x, y, z])                    # the given vector field
7  exam_DivThm_sphere(r1, X, a)
```

Divergence of the vector field $= 3$

Integrand of the domain integral $= 3r^2 \sin(\theta)$

Total flux of the vector field across the sphere surface $= 4\pi a^3$

Total divergence over the sphere $= 4\pi a^3$

Does the divergence theorem hold? True

```
1  # Consider r2:
2  r2 = Matrix([x,y,z+c/2])                    # the given vector field
3  exam_DivThm_sphere(r2, X, a)
```

Divergence of the vector field $= 3$

Integrand of the domain integral $= 3r^2 \sin(\theta)$

Total flux of the vector field across the sphere surface $= 4\pi a^3$

Total divergence over the sphere $= 4\pi a^3$

Does the divergence theorem hold? True

```
1  # Consider r3:
2  r3 = Matrix([x,y,z+2*c])                     # the given vector field
3  exam_DivThm_sphere(r3, X, a)
```

Divergence of the vector field $= 3$

Integrand of the domain integral $= 3r^2 \sin(\theta)$

Total flux of the vector field across the sphere surface $= 4\pi a^3$

Total divergence over the sphere $= 4\pi a^3$

Does the divergence theorem hold? True

We found the following:

1. The divergence of the position (radial flow) field is 3.
2. The total divergence over a sphere is three times the sphere's volume. This finding can be used to compute the volume of a domain similar to how we computed the area of a 2D domain.
3. The flux over the closed sphere equals the total divergence, and the divergence theorem holds.

This finding is also true for $\mathbf{r}_2$ and $\mathbf{r}_3$, implying that the center of the radial flow is immaterial in these cases. This is because the construction of the position field results in the same constant divergence of 3 throughout the entire space. The shifting constant is eliminated by the differentiation in the divergence operator. The total divergence of any domain is three times the volume of the domain regardless of its shape.

6.2.4 *Example: Examination of a 3D gravity field*

Consider the 3D **gravity field** defined as

$$\mathbf{r}_g = \begin{bmatrix} -x/r^3 \\ -y/r^3 \\ -z/r^3 \end{bmatrix} \tag{6.6}$$

where r is given by

$$r = \sqrt{x^2 + y^2 + z^2} \tag{6.7}$$

Check whether the surface integral over the sphere centered at the origin equals the volume integral over the same sphere and whether the divergence theorem holds.

Solution: We use the same code to perform the task:

```
1  x, y, z = symbols('x, y, z', real=True)
2  a = symbols('a', positive=True)              # radius of the sphere
3  r_ = sp.sqrt(x**2 + y**2 + z**2)
4  V = Matrix([-x/r_**3, -y/r_**3, -z/r_**3])    # the gravity fields
5  exam_DivThm_sphere(V, X, a)
```

Divergence of the vector field $= 0$

Integrand of the domain integral $= 0$

Total flux of the vector field across the sphere surface $= -4\pi$

Total divergence over the sphere $= 0$

Does the divergence theorem hold? False

This time, it is observed that the divergence theorem does not hold for this case. Although the evaluation of the flux over the sphere's surface is correct, the issue arises with the computation of the divergence of the gravitational field. Since r^3 is in the denominator of the vector field functions, there is a singular point, also referred to as a pole, at $(0,0,0)$. The field functions are not differentiable there. In the **exam_DivThm_sphere**() function, we computed its divergence anyway, which resulted in a false zero! This result is incorrect. The condition set for the divergence theorem in Eq. (6.3) is violated. The divergence theorem is not supposed to hold, and the code reflects this. Note that the effect of the pole is captured by the surface integral. This case is quite similar to the singular position and spin fields in 2D discussed in Chapter 4.

The same issue arises for the electrostatic field defined as $[x/r^3, y/r^3, z/r^3]^\top$ (differing by a sign). Similar issues often arise in wave propagation problems, where integrals using complex contours are required to handle such cases. Since the solution can be multiple for domains with poles, one must consider the physical context of the problem to determine the appropriate solution that makes physical sense. Interested readers may refer to Ref. [5]. We will not discuss this further in this volume but make note of the following:

> The divergence theorem requires that the vector field be differentiable.

6.2.5 *Example: Examination of heat flux vector in the Sun*

Assume the Sun is an ideal spherical ball and the temperature inside is point symmetric with respect to its center and can be given as

$$T = \ln\frac{1}{r} \tag{6.8}$$

where r is the radial coordinate: $r = \sqrt{x^2 + y^2 + z^2}$.

Based on Fourier's law, the heat flux vector is given as

$$\mathbf{q} = -\nabla T \tag{6.9}$$

Check whether the surface integral of the heat flux vector $\mathbf{q}$ over the sphere's surface equals the volume integral of its divergence over the same sphere and whether the divergence theorem holds.

Solution: We use the same code to perform the task:

```
1  r, x, y, z = symbols('r, x, y, z', real=True)
2  a = symbols('a', positive=True)              # radius of the sphere
3  X = Matrix([x, y, z])
4  r_ = sp.sqrt(x**2 + y**2 + z**2)
5
6  T = sp.log(1/r_)
7  q = -gr.grad_f(T, X)
8  display(Math(f" \\ q=-∇T = {latex(q)} "))
9  exam_DivThm_sphere(q, X, a)
```

$$\mathbf{q} = -\nabla T = \begin{bmatrix} \frac{x}{x^2+y^2+z^2} \\ \frac{y}{x^2+y^2+z^2} \\ \frac{z}{x^2+y^2+z^2} \end{bmatrix}$$

Divergence of the vector field $= \dfrac{1}{x^2 + y^2 + z^2}$

Integrand of the domain integral $= \sin(\theta)$

Total flux of the vector field across the sphere surface $= 4\pi a$

Total divergence over the sphere $= 4\pi a$

Does the divergence theorem hold? True

Note that in this case, the temperature is singular at $r = 0$ and the heat flux vector $\mathbf{q}$ is also not defined at $r = 0$. However, this singularity is removable because the sum of the numerators of the components of $\mathbf{q}$ approaches zero at the same rate as the denominator [4]. The divergence (the sum of the partial derivatives) becomes $\frac{1}{x^2+y^2+z^2} = \frac{1}{r^2}$ (see the result given above), which is integrable in a spherical domain. The volume scalar (see Ref. [4]) is $r^2 \sin\theta$, which cancels out $1/r^2$, effectively removing the singularity and hence can be integrated. Due to this feature, the divergence theorem holds for this heat flux vector field.

6.3 Application: Derivation of PDEs

Another important application of the divergence theorem is to derive partial differential equations (PDEs) that govern significant real-life problems. As an example, this section presents a procedure that leads to one PDE for the mass conservation law, which is one of the most fundamental laws.

6.3.1 *Physical law in integral form*

Many laws in physics can often be easily written in integral form. For example, the mass conservation law can be stated as

$$\underbrace{\oint_S \rho\mathbf{F} \cdot \mathbf{n}dS}_{\text{mass leaves along } S} = -\underbrace{\frac{d}{dt} \iiint_{\mathcal{D}} \rho dV}_{\text{change in mass over } \mathcal{D}} \tag{6.10}$$

This is a physical statement of mass conservation. The amount of mass leaving the surface is equal to the negative rate of change of mass within the volume. It represents the integral form of the physical mass conservation law.

Using the divergence theorem in Eq. (6.3), the left side of the foregoing equation can be written as

$$\oint_S \rho\mathbf{F} \cdot \mathbf{n}dS = \iiint_{\mathcal{D}} \underbrace{\nabla \cdot (\rho\mathbf{F})}_{\text{div}(\rho\mathbf{F})} dV \tag{6.11}$$

Equations (6.10) and (6.11) give

$$\iiint_{\mathcal{D}} \nabla \cdot (\rho \mathbf{F}) \mathrm{d}V = -\frac{\mathrm{d}}{\mathrm{d}t} \iiint_{\mathcal{D}} \rho \mathrm{d}V \tag{6.12}$$

Rearranging it, we have

$$\iiint_{\mathcal{D}} \left[\frac{\mathrm{d}\rho}{\mathrm{d}t} + \nabla \cdot (\rho \mathbf{F}) \right] \mathrm{d}V = 0 \tag{6.13}$$

6.4 Physical law in PDEs

Now, since the domain $\mathcal{D}$ in the foregoing equation is arbitrary — implying that it can be of any size and shape — Eq. (6.13) must hold universally. This can only happen if the term in the square brackets is zero. Conversely, if the squared term is nonzero in some domain, we can always find a smaller domain within that domain where the squared term is either all positive or all negative, which would lead to a nonzero integral. Therefore, for Eq. (6.13) to hold for any $\mathcal{D}$, the squared term must be zero. We finally arrive at

$$\frac{\partial \rho}{\partial t} + \nabla \cdot (\rho \mathbf{F}) = 0 \tag{6.14}$$

This is the PDE for the mass conservation law, commonly referred to as the mass continuity equation. We can see that the divergence theorem is a useful tool for deriving governing PDEs for physical problems, provided they have a corresponding physical law in integral form. It is applicable in deriving PDEs for various physical phenomena, including heat transfer and solid mechanics problems [3].

Why do we need a PDE? The reason we require a PDE for a problem is that effective computational methods have been developed to solve them and find useful solutions. Notably, this volume will not delve into techniques for solving PDEs as we plan to dedicate a future volume to this topic.

6.5 Concluding remarks

This chapter focused on volume integrals of 3D vector functions. We presented formulas, techniques, and Python codes for computing the volume of a 3D body and the total divergence of a vector function over a 3D domain. We conclude the chapter with the following remarks:

1. The divergence theorem states that the total divergence of a vector field over a domain equals the flux of the vector field across the surface of that domain. It serves as a mathematical description of the law of mass conservation in 3D.
2. The divergence theorem was examined using spherical domains for various types of vector fields.
3. The divergence theorem is instrumental in deriving PDEs for many problems.

The following chapter discusses Stokes' theorem, which is the 3D extension of Green's circulation theorem and represents a law of energy conservation in 3D space.

References

[1] G. Strang, *Calculus*, Wellesley-Cambridge Press, Massachusetts, 1991.
[2] G.R. Liu, *Numbers and Functions: Theory, Formulation, and Python Codes*, World Scientific, New Jersey, 2024.
[3] G.R. Liu, *Mechanics of Materials: Formulations and Solutions with Python*, World Scientific, New Jersey, 2024.
[4] G.R. Liu, *Calculus: A Practical Course with Python*, World Scientific, New Jersey, 2025.
[5] G.R. Liu and Z.C. Xi, *Elastic Waves in Anisotropic Laminates*, CRC Press, New York, 2001.

Chapter 7

Stokes' Theorem

```
 1  # Place cursor in this cell, and press Ctrl+Enter to import dependences.
 2  import sys                        # For accessing the computer system
 3  sys.path.append('../grbin/')          # Add in the path to your system
 4
 5  from commonImports import *      # Import dependences from '../grbin/'
 6  import grcodes as gr               # Import the module of the author
 7  importlib.reload(gr)            # When grcodes is modified, reload it
 8
 9  init_printing(use_unicode=True)       # For Latex-like quality printing
10  np.set_printoptions(precision=4,suppress=True,
11                      formatter={'float_kind': '{:.4e}'.format})
```

In Chapter 4, Green's theorem for two-dimensional vector fields is discussed in detail, involving curve integrals and 2D domain integrals. This chapter studies Stokes' theorem, which is a three-dimensional (3D) extension of Green's circulation theorem, leading to a law of energy conservation over a surface in 3D space. The focus of this chapter will be on techniques and code for examining Stokes' theorem. Techniques and codes will also be presented to generate potential functions for conservative fields.

This chapter is written with reference to textbooks in Refs. [1–3]. Wikipedia pages, particularly those on integral (https://en.wikipedia.org/wiki/Integral), surface integral (https://en.wikipedia.org/wiki/Surface_integral), Green's theorem (https://en.wikipedia.org/wiki/Green%27s_theorem), and Stokes' theorem (https://en.wikipedia.org/wiki/Stokes%27_theorem), serve as valuable additional references.

Both NumPy and SymPy are used in the development of code for the demonstration examples. Discussions with ChatGPT, Gemini, and Bing have also greatly helped in coding and in the preparation of this chapter.

7.1 Formulation

Consider a differentiable vector field $\mathbf{V}$ defined on a simply connected surface $\mathcal{S}$ in 3D space bounded by a smooth (or at least piecewise smooth with finite pieces) curve $\mathcal{C}$. It is expressed as

$$\mathbf{V} = \begin{bmatrix} V_x(x,y,z) \\ V_y(x,y,z) \\ V_z(x,y,z) \end{bmatrix} = \begin{bmatrix} M(x,y,z) \\ N(x,y,z) \\ P(x,y,z) \end{bmatrix} \tag{7.1}$$

where V_x, V_y, and V_z are the component functions of $\mathbf{V}$. These are often denoted in the literature as M, N, and P, respectively.

Stokes' theorem states that

$$\underbrace{\oint_{\mathcal{C}} (V_x \, dx + V_y \, dy + V_z \, dz)}_{\text{total work along curve } \mathcal{C}}$$

$$= \underbrace{\iint_{\mathcal{S}} \left[\left(\frac{\partial V_z}{\partial y} - \frac{\partial V_y}{\partial z} \right) \underbrace{dydz}_{n_x dS} + \left(\frac{\partial V_x}{\partial z} - \frac{\partial V_z}{\partial x} \right) \underbrace{dxdz}_{n_y dS} + \left(\frac{\partial V_y}{\partial x} - \frac{\partial V_x}{\partial y} \right) \underbrace{dxdy}_{n_z dS} \right]}_{\text{total curl on surface } \mathcal{S}}$$

$$\tag{7.2}$$

or simply

$$\oint_{\mathcal{C}} \mathbf{V} \cdot \mathbf{t} ds = \iint_{\mathcal{S}} (\text{curl } \mathbf{V}) \cdot \mathbf{n} dS \tag{7.3}$$

where curl $\mathbf{V}$ is defined in Eq. (2.21) and $\mathbf{n} dS = d\mathbf{r}$ where $\mathbf{r}$ is the position vector (see Chapter 4).

Here, we also require that the surface $\mathcal{S}$ be simply connected, as discussed in Section 4.1.1 of Chapter 4. If the domain is multiply connected, cuts are needed to make it simply connected.

In general, we also require that all these integrands be integrable. Since the vector function is assumed to be differentiable, these integrands are usually (Riemann) integrable although we may not be able to find the analytical form of antiderivatives.

In Eq. (7.3), both $\mathcal{C}$ and $\mathcal{S}$ are generally in 3D space, as shown in Fig. 7.1.

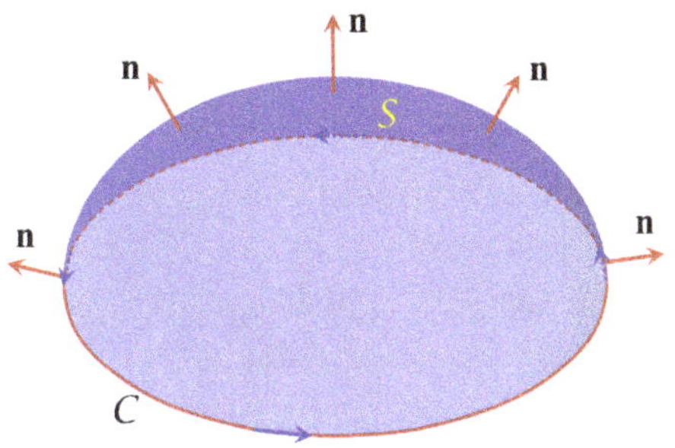

Figure 7.1. A surface $\mathcal{S}$ in 3D space bounded by an edge curve $\mathcal{C}$.

The curve integral follows the standard counterclockwise (CCW) and right-hand rule convention. However, since everything is in 3D, we need to be more careful: When walking along $\mathcal{C}$, one must always keep $\mathcal{S}$ on the left side and ensure that one's head is oriented in the same direction as $\mathbf{n}$, the unit normal of $\mathcal{S}$. Let's refer to this as the CCW right-hand head-up rule for convenience.

The following code defines in Python the vectors used in Eq. (7.2):

```python
1  # Define variables and functions:
2
3  x, y, z = sp.symbols('x, y, z ')
4  X = sp.Matrix([x, y, z])
5  nx, ny, nz = sp.symbols('n_x, n_y, n_z')
6
7  Vx = Function('V_x')(x, y, z)
8  Vy = Function('V_y')(x, y, z)
9  Vz = Function('V_z')(x, y, z)
10 V = sp.Matrix([Vx, Vy, Vz])
11
12 V = sp.Matrix([Vx, Vy, Vz])
13 n = sp.Matrix([nx, ny, nz])
14 curl_V = gr.curl_vf(V, X)
15 display(Math(f" \\text{{curl_V = }}{latex(curl_V)};\\;\\;\
16                \\text{{n on surface $S$ = }}{latex(n)}"))
```

$$\text{curl\_V} = \begin{bmatrix} -\frac{\partial}{\partial z}V_y(x,y,z) + \frac{\partial}{\partial y}V_z(x,y,z) \\ \frac{\partial}{\partial z}V_x(x,y,z) - \frac{\partial}{\partial x}V_z(x,y,z) \\ -\frac{\partial}{\partial y}V_x(x,y,z) + \frac{\partial}{\partial x}V_y(x,y,z) \end{bmatrix} ; \quad \mathbf{n} \text{ on surface } S = \begin{bmatrix} n_x \\ n_y \\ n_z \end{bmatrix}$$

```python
1  # curl_V.dot(n) # one may take a look at the integrand, long expression
```

7.2 Comparison with the Green's theorem

To make sense of the foregoing equation, let us compare it with Green's theorem along a curve in a 2D domain:

$$\oint_{\mathcal{C}} \mathbf{V} \cdot \mathbf{t}\, \mathrm{d}s = \iint_{\mathcal{D}} \mathrm{curl}_z\, \mathbf{V}\, \mathrm{d}A \tag{7.4}$$

where $\mathrm{curl}_z \mathbf{V}$ is the 2D curl, the z-component of the 3D curl defined in Eq. (2.22), and hence is a scalar.

The major differences are as follows:

1. The $\mathcal{C}$ in Eq. (7.3) is a curve in 3D space, whereas the $\mathcal{C}$ in Eq. (7.4) is a curve in 2D space.
2. The $\mathcal{S}$ in Eq. (7.3) is a surface in 3D space, whereas $\mathcal{D}$ in Eq. (7.4) is a 2D plane.

A more detailed comparison of Eq. (7.4) with Eq. (7.3) can help in better understanding of Stokes' theorem:

1. Assume $\mathcal{S}$ lies in the (x, y) plane, in which case $\mathbf{n}$ points in the z-direction. The left sides of these two equations are clearly the same. Additionally, the 3D curl of $\mathbf{V}$ projected onto $\mathbf{n}$, the dot product on the right side of Eq. (7.4), becomes $\mathrm{curl}_z \mathbf{V}$. This implies that the right sides of these two equations are also the same, making Eq. (7.3) a special case of Eq. (7.4). Both equations state that the work done by the force field $\mathbf{V}$ along the closed curve $\mathcal{C}$ equals the total circulation of $\mathbf{V}$ over the surface $\mathcal{S}$ with respect to the z-axis.
2. In 3D cases, we use Eq. (7.3). The measure of work done on the left side of Eq. (7.4) should still use the dot product, but along a 3D curve $\mathcal{C}$. The measure of the local circulation over a differential area of the 3D surface $\mathcal{S}$ should be the local curl projected onto the local normal $\mathbf{n}$, which varies from location to location. The total circulation is the integral of all these local circulations. Thus, Eq. (7.3) provides a correct measure of a conservation law for a surface in 3D space.

More detailed proofs can be found in Ref. [1] and on the Wikipedia page on Stokes' theorem (https://en.wikipedia.org/wiki/Stokes%27_theorem).

Let us look at some examples using Python. First, we will write the following code to examine whether a given vector field satisfies Stokes' theorem:

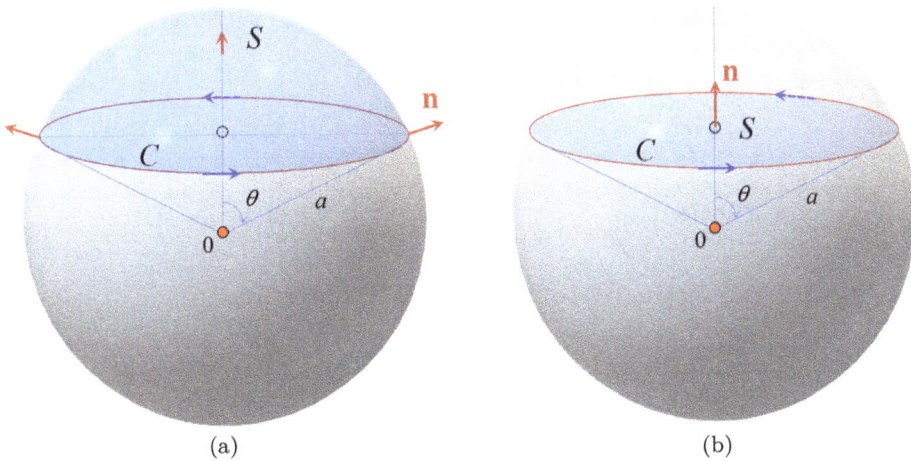

(a)   (b)

Figure 7.2. (a) A spherical surface (shaded) bounded by a circle (red) and (b) a flat disk (shaded) bounded by the same circle (red).

7.3 Python function for the examination of Stokes' theorem

Let us write the following Python code to examine Stokes' theorem for any given vector field. This code will help us study multiple cases. To facilitate the creation of surfaces and their boundary curves, we use the upper part of a sphere to create a partial spherical surface S with its circular boundary curve C, as illustrated in Fig. 7.2(a).

The tasks that the Python function can perform are detailed in the docstring of the code:

```python
def exam_StokesThm_sphere(V, X, a, polarθ):
    '''Examine a given vector field V that is differentiable over
       a spherical domain. The code prints out
       1 curl vector of the vector field V.
       2 The total circulation of the field over a partial surface of
         a sphere controled by polarθ.
       3 The total work done by V along the circular boundary curve of
         the partial surface.
       4 Whether or not the Stokes' theorem holds.
       The code uses a upper part of a sphere to create a partial
       spheric surface S with a circular boundary curve C. The size of
       the surface and curve is controled by the polar angle: polar θ.
       Positive n is outwords from the sphere.
    '''
```

```
15      θ, φ = symbols('θ, φ', real=True)
16      x, y, z = X
17
18      # Integral over the closed base circle C:
19      θ_ = polarθ
20      xC = a*sin(θ_)*cos(φ)              # xC, yC, and zC are x,y,z on curve
21      yC = a*sin(θ_)*sin(φ)
22      zC = a*cos(θ_)
23      dicC = {x:xC, y:yC, z:zC}
24
25      dr = Matrix([xC.diff(φ), yC.diff(φ), zC.diff(φ)])
26      fIC = V.T@dr               # dr = ndS integrand for the curve integral
27      Int = integrate
28      Fdr = Int(fIC.subs(dicC), (φ,0,2*sp.pi)).simplify()
29
30      # Integral over the partial spherical surface S:
31      xS = a*sin(θ)*cos(φ)        # xS, yS, and zS are x,y,z on the surface
32      yS = a*sin(θ)*sin(φ)
33      zS = a*cos(θ)
34      dicS = {x:xS, y:yS, z:zS}
35      curl_V = gr.curl_vf(V, X)
36      display(Math(f" \\text{{Curl.T of the vector field}} = \
37                                      {latex(curl_V.T)}"))
38      fIS = (curl_V.T@X/sp.sqrt(x**2+y**2+z**2))*(a**2 *sin(θ)).simplify()
39      # integrand for the surface integral, detJ = a**2 *sin(θ)
40      curlFS = Int(Int(fIS.subs(dicS),(φ,0,2*sp.pi)),(θ,0,θ_)).simplify()
41
42      Fdr = Fdr.extract([0],[0])[0,0] if Fdr else Fdr # for print results
43      curlFS = curlFS.extract([0],[0])[0,0] if curlFS else curlFS
44      display(Math(f" \\text{{Work done by the field along base circle}}\
45                                      = {latex(Fdr)}"))
46      display(Math(f" \\text{{Total circulation over spheric surface}} \
47                                      = {latex(curlFS)}"))
48      print(f" Does the Stokes' theorem hold? {Fdr==curlFS}")
```

7.3.1 *Example: Spin fields, circulation, and Stokes' theorem*

Consider a spin field given by Eq. (3.7). These fields represent 3D spin effects with respect to the x-, y-, and z-axes. The surface S is the upper part of a sphere, and the curve C is the circular boundary of S, as shown in Fig. 7.2.

Tasks:

1. Find the work done by the vector field via curve integration along C.
2. Show that the spin field has a curl.
3. Find the total circulation of the field via surface integration using the curl.

4. Show that Stokes' theorem holds: The total circulation obtained by the surface integral of the curl over any part of the spherical surface $\mathcal{S}$ should be the same as the work done obtained by the curve integral of the vector field along $\mathcal{C}$.

Solution: We use `exam_StokesThm_sphere()` and the following snippets to perform the task. Different polar angles θ are used to create surfaces (and curves) of various sizes:

```
1  # Define variables:
2  x, y, z = symbols('x, y, z', real=True)
3  X = Matrix([x, y, z])
4  a = symbols('a', positive=True)              # Radius of the sphere
5
6  sx = Matrix([0, -z, y])                       # the given vector fields
7  sy = Matrix([z, 0, -x])
8  sz = Matrix([-y, x, 0])
9  exam_StokesThm_sphere(sz, X, a, sp.pi/2)      # use upper half sphere
```

Curl.T of the vector field $= \begin{bmatrix} 0 & 0 & 2 \end{bmatrix}$

Work done by the field along base circle $= 2\pi a^2$

Total circulation over spheric surface $= 2\pi a^2$

Does the Stokes' theorem hold? True

It is found that the spin field s_z has a circulation. The total circulation computed using the surface integral over the surface is the same as the work done along the curve, confirming that Stokes' theorem holds. Note also that the total circulation is proportional to the area projected onto the x–y plane. If the projected area is reduced by changing the polar angle, the total circulation should decrease proportionally. The following snippet confirms this:

```
1  exam_StokesThm_sphere(sz, X, a, sp.pi/3)      # use upper 1/3 sphere
2  #exam_StokesThm_sphere(sz, X, a, 2.*sp.pi/3)  # or upper 2/3 sphere
```

Curl.T of the vector field $= \begin{bmatrix} 0 & 0 & 2 \end{bmatrix}$

Work done by the field along base circle $= \dfrac{3\pi a^2}{2}$

Total circulation over spheric surface $= \dfrac{3\pi a^2}{2}$

Does the Stokes' theorem hold? True

Let us use the semi-sphere to check on $\mathbf{s}_y$:

```
1  exam_StokesThm_sphere(sy, X, a, sp.pi/2)
```

Curl.T of the vector field $= \begin{bmatrix} 0 & 2 & 0 \end{bmatrix}$

Work done by the field along base circle $= 0$

Total circulation over spheric surface $= 0$

```
Does the Stokes' theorem hold? True
```

This time, the total circulation is zero although the curl of the field is nonzero. This is because this spin field is circulating with respect to the y-axis, while the second component is nonzero. Any surface formed by the upper portion of a sphere cannot capture the circulation of such a spin field. If a properly horizontally oriented partial sphere is used, the circulation should be captured. The same applies to $\mathbf{s}_x$. In any case, Stokes' theorem holds. We note the following:

> Stokes' theorem captures the spin behavior of a vector field. It is a theorem about the curl of a vector field. Since the curl of any vector field is a vector, the surface must have the right orientation to capture the spin behavior in addition to the size of the surface.

7.3.2 *Example: Position fields, noncirculative*

Consider position fields defined in Eq. (3.2); by using the same Python function, we can easily examine the circulation:

```
1  r1 = Matrix([x, y, z])                        # the given vector field
2  exam_StokesThm_sphere(r1, X, a, sp.pi/3)
```

Curl.T of the vector field $= \begin{bmatrix} 0 & 0 & 0 \end{bmatrix}$

Work done by the field along base circle $= 0$

Total circulation over spheric surface $= 0$

```
Does the Stokes' theorem hold? True
```

We found zero circulation. This is because the position (radial) field is a circulation-free field. The right side of Stokes' theorem is always zero because the integrand (the curl) is zero. Hence, this holds true regardless of the shape of the surface used. The left side is not obviously zero, but Stokes' theorem states that it must be.

This is also true for the other two radial fields or any radial field with its center shifted (readers may try it using the same code) because any radial field is not circulating. We note the following:

> Stokes' theorem does not capture any radial behavior. It is a theorem about the curl of a vector field.

7.3.3 *Example: Stokes' theorem on a gravity field*

Consider the 3D gravity field given in Eq. (6.6). Let us conduct the same examination as the previous example using the same Python function:

```
1  x, y, z = symbols('x, y, z', real=True)
2  r = sp.sqrt(x**2 + y**2 + z**2)
3  V = Matrix([-x/r**3, -y/r**3, -z/r**3])          # the gravity fields
4  exam_StokesThm_sphere(V, X, a, sp.pi/3)
```

Curl.T of the vector field $= \begin{bmatrix} 0 & 0 & 0 \end{bmatrix}$

Work done by the field along base circle $= 0$

Total circulation over spheric surface $= 0$

Does the Stokes' theorem hold? True

We found zero circulation, and it is also a curl-free field. This is expected because a gravity field is not circulating. We note the following:

> The circulation of a gravity field will always be zero regardless of the shape of the surface used.

7.3.4 *Example: Stokes' theorem on arbitrary vector fields*

Consider an arbitrarily constructed vector field:

$$\mathbf{V} = \begin{bmatrix} -xy \\ xz \\ zx \end{bmatrix} \tag{7.5}$$

The snippet to examine it is as follows:

```
1  Vx = -x**2*y+x; Vy = x*z*y; Vz = z*x+y
2  V = Matrix([Vx, Vy, Vz])                          # an arbitrary field
3  exam_StokesThm_sphere(V, X, a, sp.pi/3)
```

Curl.T of the vector field $= \begin{bmatrix} -xy + 1 & -z & x^2 + yz \end{bmatrix}$

Work done by the field along base circle $= \dfrac{9\pi a^4}{64}$

Total circulation over spheric surface $= \dfrac{9\pi a^4}{64}$

```
Does the Stokes' theorem hold? True
```

The curl of this vector field has all nonzero components. Stokes' theorem still holds; it captures the circulation of the field. Let us check its divergence:

```
1  a = symbols('a', positive=True)          # radius of the sphere
2  gr.exam_DivThm_sphere(V, X, a)     # the divergence of the vector field
```

Divergence of the vector field $= -2xy + xz + x + 1$

Total flux of the vector field across the sphere surface $= \dfrac{4\pi a^3}{3}$

Total divergence over the sphere $= \dfrac{4\pi a^3}{3}$

```
Does the divergence theorem hold? True
```

It has divergence, and the divergence theorem also holds. It captures the flux of the vector field across the spherical surface.

This example shows the following:

> The divergence theorem and Stokes' theorem capture different effects of a vector field. Different fields with different measures have different effects.

7.3.5 *Example: Surfaces with the same boundary curve*

Consider now two different surfaces, both with the same boundary curve: one is a spherical surface and the other is a flat disk, as shown in Fig. 7.2.

Show the following:

1. For the arbitrary vector field given in Eq. (7.5), Stokes' theorem holds for both surfaces: The total circulation over these surfaces equals the work done along the same boundary curve.
2. The same holds for a spin field.

This time, we need to write the following code to perform this task. The code is largely the same as **exam_StokesThm_sphere()**. The difference is the use of a flat disk as the surface, for which the surface integral requires a new variable r.

```
1  def exam_StokesThm_circle(V, X, a, polarθ):
2      '''Perform the same tasks as exam_StokesThm_sphere(), but the surface
3        is a disk.
4      '''
5      r = symbols('r', nonnegative=True)  # radial coordinate for the disk
6      θ, φ = symbols('θ, φ', real=True)
7      x, y, z = X
8
9      # Integral over the closed circular curve C:
10     θ_ = polarθ
11     xC = a*sin(θ_)*cos(φ)          # xC, yC, and zC are x,y,z on curve C
12     yC = a*sin(θ_)*sin(φ)
13     zC = a*cos(θ_)
14     dicC = {x:xC, y:yC, z:zC}
15
16     dr = Matrix([xC.diff(φ), yC.diff(φ), zC.diff(φ)])
17     fIC = V.T@dr                   # dr = ndS integrand for the curve integral
18     Int = integrate
19     Fdr = Int(fIC.subs(dicC), (φ,0,2*sp.pi)).simplify()
20
21     # Integral over the disk S:
22     xS = r*cos(φ)                  # xS, yS, and zS are x,y,z on the disk surface
23     yS = r*sin(φ)
24     zS = a*cos(θ_)
25     dicS = {x:xS, y:yS, z:zS}
26     curl_V = gr.curl_vf(V, X)
27     display(Math(f" \\text{{Curl.T of the vector field in the disk}} =\
28                               {latex(curl_V.T)}"))
29     n = Matrix([0, 0, 1])
30     fIS = (curl_V.T@n)*r           # integrand over the disk, # use detJ = r
31     a_ = a*sin(θ_)
32     curlFS =Int(Int(fIS.subs(dicS),(φ,0,2*sp.pi)), (r,0,a_)).simplify()
33
34     Fdr = Fdr.extract([0],[0])[0,0] if Fdr else Fdr # for print results
35     curlFS = curlFS.extract([0],[0])[0,0] if curlFS else curlFS
36     display(Math(f" \\text{{Work done by the field along curve C}} \
37                               = {latex(Fdr)}"))
38     display(Math(f" \\text{{Total circulation over disk surface}} \
39                               = {latex(curlFS)}"))
40     print(f" Does the Stokes' theorem hold? {Fdr==curlFS}")
```

The results using the flat disk for the vector field given in Eq. (7.5) are as follows:

```
1  exam_StokesThm_circle(V, X, a, sp.pi/4)          # use upper 1/4 sphere
```

$\text{Curl.T of the vector field in the disk} = \begin{bmatrix} -xy+1 & -z & x^2+yz \end{bmatrix}$

$\text{Work done by the field along curve C} = \dfrac{\pi a^4}{16}$

Total circulation over disk surface $= \dfrac{\pi a^4}{16}$

```
Does the Stokes' theorem hold? True
```

The results using the spherical surface for the vector field given in Eq. (7.5) are as follows:

```
1  exam_StokesThm_sphere(V, X, a, sp.pi/4)        # use upper 1/4 sphere
```

Curl.T of the vector field $= \begin{bmatrix} -xy+1 & -z & x^2+yz \end{bmatrix}$

Work done by the field along base circle $= \dfrac{\pi a^4}{16}$

Total circulation over spheric surface $= \dfrac{\pi a^4}{16}$

```
Does the Stokes' theorem hold? True
```

As observed, the results obtained using the flat circular surface and the spherical surface are the same.

Next, let us consider the spin field s_z. First, we use the flat disk:

```
1  exam_StokesThm_circle(sz, X, a, sp.pi/4)       # use upper 1/4 sphere
```

Curl.T of the vector field in the disk $= \begin{bmatrix} 0 & 0 & 2 \end{bmatrix}$

Work done by the field along curve $C = \pi a^2$

Total circulation over disk surface $= \pi a^2$

```
Does the Stokes' theorem hold? True
```

Now, use the spherical surface for the same spin field:

```
1  exam_StokesThm_sphere(sz, X, a, sp.pi/4)       # use upper 1/4 sphere
```

Curl.T of the vector field $= \begin{bmatrix} 0 & 0 & 2 \end{bmatrix}$

Work done by the field along base circle $= \pi a^2$

Total circulation over spheric surface $= \pi a^2$

```
Does the Stokes' theorem hold? True
```

We note the following:

> Stokes' theorem holds for differentiable vector fields over any (piecewise) smooth surface regardless of shape and size.

7.4 Conservative fields and potential functions in 3D

In science and engineering, some special and important field functions describe the behaviors of commonly encountered problems. Conservative fields are among these, and we studied them earlier for 2D cases. Let us extend the study to 3D cases.

Consider a vector field $\mathbf{V} = [V_x(x, y, z) \quad V_y(x, y, z) \quad V_z(x, y, z)]^\top$. If it is conservative, it has the following properties:

1. A continuously differentiable potential function ϕ can be found such that $V_x = \frac{\partial \phi}{\partial x}$, $V_y = \frac{\partial \phi}{\partial y}$, and $V_z = \frac{\partial \phi}{\partial z}$. This means that it is a gradient field: Its component functions are the components of the gradient of a potential function.
2. It is curl-free: $\frac{\partial V_z}{\partial y} = \frac{\partial V_y}{\partial z}$, $\frac{\partial V_x}{\partial z} = \frac{\partial V_z}{\partial x}$, and $\frac{\partial V_y}{\partial x} = \frac{\partial V_x}{\partial y}$. This means that the circulation over any surface is zero. This can be easily confirmed using item 1 via partial differentiation.
3. The work done around any closed path is zero: $\oint \mathbf{V} \cdot d\mathbf{r} = 0$. This is a consequence of item 2 and Stokes' theorem.
4. The work done along any curve $\int_C \mathbf{V} \cdot d\mathbf{r}$ depends only on the starting point (a) and the ending point (b), not on the path chosen. This is the **fundamental theorem for a line integral**.

A vector field $\mathbf{V}$ possessing any one of these properties has all these properties. These properties can then be used to find the potential function ϕ. Note that any constant in the potential function ϕ has no effect in terms of computing the components of the vector function because all the vector field components depend only on their derivatives.

We write the following Python function to automatically find the potential function of a conservative field. It is an extension of the 2D code developed earlier and is self-explanatory:

```
1  def find_potential_f(Vx, Vy, Vz):
2      '''
3          Find the potential function φ for a given 3D vector field
4          component functions, Vx, Vy, Vz. The field must be curl-free.
5          Vz needs not to provide. It will be computed after φ is found.
6      '''
7      # Use condition: ∂φ/∂x = Vx
8      C = sp.Function('C')(y)            # define symbolic integral functions
9      φC = integrate(Vx, (x)) + C        # φ with C(y); used Vx = ∂φ/∂x
10
```

```
11      # use condition: Vy = ∂φ/∂y
12      eqn = φC.diff(y)- Vy                           # create an equation
13      sln_C = sp.dsolve(eqn, C)              # solve the eqn for C(y)
14
15      φC = φC.subs(C, sln_C.args[1])    # put solution of C(y) back to φC
16      φ  = φC.subs(φC.args[0],0).simplify() # set integral constant to 0
17      display(Math(f"\\text{{Potential function φ found = }}{latex(φ)}"))
18      Vz = φ.diff(z).simplify()                     # Compute Vz(x,y,z)
19
20      # Check all conditions:
21      gr.printx('(φ.diff(x)-Vx).simplify()') # check condition: Vx=∂φ/∂x
22      gr.printx('(φ.diff(y)-Vy).simplify()') # check condition: Vy=∂φ/∂y
23      gr.printx('(φ.diff(z)-Vz).simplify()') # check condition: Vz=∂φ/∂z
24
25      return φ, Vz
```

7.4.1 *Example: Conservation tests and potential functions*

Consider a vector field given by

$$\mathbf{V} = \begin{bmatrix} 2xy + y^2 \\ x^2 + 2xy + z^2 \\ 2yz \end{bmatrix} \tag{7.6}$$

1. Determine whether this field is conservative.
2. If it is, find the potential function.

We write the following snippet to demonstrate the examination process and find the potential function ϕ systematically using SymPy:

```
1  # Define the variables and vector functions:
2  x, y, z = symbols('x, y, z', real=True)
3  Vx, Vy, Vz = symbols('V_x, V_y, V_z', real=True)
4
5  Vx = 2*x*y + y**2
6  Vy = x**2 + 2*x*y + z**2
7  Vz = 2*y*z
8
9  # Check the curl of the vector field before proceeding:
10 gr.printx('Vz.diff(y)-Vy.diff(z)')
11 gr.printx('Vx.diff(z)-Vz.diff(x)')
12 gr.printx('Vy.diff(x)-Vx.diff(y)')
```

```
Vz.diff(y)-Vy.diff(z) = 0
Vx.diff(z)-Vz.diff(x) = 0
Vy.diff(x)-Vx.diff(y) = 0
```

The condition in item 2 is satisfied, so there exists a potential function ϕ. We can then use other conditions to determine ϕ. Let's find it using our Python function `find_potential_f()`:

```
1  φ, _ = find_potential_f(Vx, Vy, Vz) # since we know Vz, we put it in
2  #display(Math(f"\\text{{Potential function φ found = }}{latex(φ)}"))
```

Potential function ϕ found $= y\left(x^2 + xy + z^2\right)$

```
(φ.diff(x)-Vx).simplify() = 0
(φ.diff(y)-Vy).simplify() = 0
(φ.diff(z)-Vz).simplify() = 0
```

We have successfully found the potential function, and all the test conditions are satisfied.

Next, let us consider another example from Ref. [1]:

$$\mathbf{V} = \begin{bmatrix} 2xy \\ x^2 + z \\ y \end{bmatrix} \tag{7.7}$$

```
1  # Define the variables and vector functions:
2  Vx = 2*x*y; Vy = x**2 + z; Vz = y
3
4  # Check the curl of the vector field before proceed:
5  gr.printx('Vz.diff(y)-Vy.diff(z)')
6  gr.printx('Vx.diff(z)-Vz.diff(x)')
7  gr.printx('Vy.diff(x)-Vx.diff(y)')
```

```
Vz.diff(y)-Vy.diff(z) = 0
Vx.diff(z)-Vz.diff(x) = 0
Vy.diff(x)-Vx.diff(y) = 0
```

```
1  φ, _= find_potential_f(Vx, Vy, Vz)  # since we know Vz, we put it in
2  #display(Math(f"\\text{{Potential function φ found = }}{latex(φ)}"))
```

Potential function ϕ found $= y\left(x^2 + z\right)$

```
(φ.diff(x)-Vx).simplify() = 0
(φ.diff(y)-Vy).simplify() = 0
(φ.diff(z)-Vz).simplify() = 0
```

7.4.2 *Example: Potential function and the third component*

Consider a vector field given by

$$\mathbf{V} = \begin{bmatrix} y^2 + y \\ 2xy + x + z^2 \\ \text{the unknown} \end{bmatrix} \tag{7.8}$$

where V_z is the unknown. Assume the vector field is conservative.

1. Find the potential function.
2. Determine the unknown component V_z.

We use the same code to complete the task:

```
1  P = Function('P')(x,y,z)         # variable for the unknown function Vz
2  Vx = y**2 + y
3  Vy = 2*x*y + x + z**2
4  Vz = P                                      # We do not know Vz
5
6  # Check the curl of the vector field before proceed:
7  gr.printx('Vz.diff(y)-Vy.diff(z)')
8  gr.printx('Vx.diff(z)-Vz.diff(x)')
9  gr.printx('Vy.diff(x)-Vx.diff(y)')
```

```
Vz.diff(y)-Vy.diff(z) = -2*z + Derivative(P(x, y, z), y)
Vx.diff(z)-Vz.diff(x) = -Derivative(P(x, y, z), x)
Vy.diff(x)-Vx.diff(y) = 0
```

It is observed that the third curl component is zero. Thus, we can proceed by using **find_potential_f()** directly to determine both ϕ and V_z:

```
1  φ, Vz = find_potential_f(Vx, Vy, Vz)              # output Vz too
2  display(Math(f"\\text{{The missing function $V_z$ found =}}{latex(Vz)}"))
3  #display(Math(f"\\text{{Potential function φ found = }}{latex(φ)}"))
```

The potential function ϕ found $= y\left(x\left(y+1\right) + z^2\right)$

```
(φ.diff(x)-Vx).simplify() = 0
(φ.diff(y)-Vy).simplify() = 0
(φ.diff(z)-Vz).simplify() = 0
```

The missing funcion V_z found $= 2yz$

We have successfully found the potential function and the third component V_z under the assumption that the vector field is conservative.

Next, let us try another example from Ref. [1]:

$$\mathbf{V} = \begin{bmatrix} yz^2 \\ xz^2 \\ \text{the unknown} \end{bmatrix} \tag{7.9}$$

```
1  # Define the variables and vector functions:
2  P = Function('Vz')(x,y,z)
3  Vx = y*z**2; Vy = x*z**2; Vz = P                    # We do not know Vz
4
5  # Check the curl of the vector field before proceed:
6  gr.printx('Vz.diff(y)-Vy.diff(z)')
7  gr.printx('Vx.diff(z)-Vz.diff(x)')
8  gr.printx('Vy.diff(x)-Vx.diff(y)')
```

```
Vz.diff(y)-Vy.diff(z) = -2*x*z + Derivative(Vz(x, y, z), y)
Vx.diff(z)-Vz.diff(x) = 2*y*z - Derivative(Vz(x, y, z), x)
Vy.diff(x)-Vx.diff(y) = 0
```

```
1  # find φ and Vz:
2
3  φ, Vz = find_potential_f(Vx, Vy, Vz)                # output Vz too
4  display(Math(f"\\text{{The missing funcion Vz found = }}{latex(Vz)}"))
5  #display(Math(f"\\text{{Potential function φ found = }}{latex(φ)}"))
```

The potential function ϕ found $= xyz^2$

```
(φ.diff(x)-Vx).simplify() = 0
(φ.diff(y)-Vy).simplify() = 0
(φ.diff(z)-Vz).simplify() = 0
```

The missing funcion Vz found $= 2xyz$

7.5 Concluding remarks

This chapter studies Stokes' theorem in detail. A number of formulas, techniques, and Python codes have been developed for examining vector function, finding potential functions and performing integrals over curves and surfaces in 3D space. We close the chapter with the following remarks:

1. Stokes' theorem measures the circulation of a vector field. It results in a $0 = 0$ equation if the field is noncirculative.
2. It is the first fundamental theorem of calculus for circulations on an enclosed surface in 3D space. The total circulation of a vector function over a surface equals the work done by the vector function along the boundary curve of the surface.

3. Using spherical surfaces is convenient for studying Stokes' theorem. Many complicated vector functions can be examined using our Python functions as long as the integrand is integrable. In other words, the domain is regular and fixed (spherical) but has certain flexibility in accommodating different types of vector functions.
4. Techniques and codes have been presented to generate potential functions as well as one of their missing components for conservative fields.

The following chapter will discuss Gauss's formula and its extensions, which are integral relations involving two functions.

References

[1]   G. Strang, *Calculus*, Wellesley-Cambridge Press, Massachusetts, 1991.
[2]   G.R. Liu, *Numbers and Functions: Theory, Formulation, and Python Codes*, World Scientific, New Jersey, 2024.
[3]   G.R. Liu, *Mechanics of Materials: Formulations and Solutions with Python*, World Scientific, New Jersey, 2024.

Chapter 8

Gauss's Formula and Beyond

```
1  # Place cursor in this cell, and press Ctrl+Enter to import dependences.
2  import sys                       # For accessing the computer system
3  sys.path.append('../grbin/')          # Add in the path to your system
4
5  from commonImports import *      # Import dependences from '../grbin/'
6  import grcodes as gr                # Import the module of the author
7  importlib.reload(gr)             # When grcodes is modified, reload it
8
9  init_printing(use_unicode=True)      # For latex-like quality printing
10 np.set_printoptions(precision=4,suppress=True,
11                     formatter={'float_kind': '{:.4e}'.format})
```

In the previous chapters, we studied several theorems, including Green's theorem, the divergence theorem, and Stokes' theorem. All these theorems involve integrals of only one function, which can be either scalar or vector. This chapter discusses Gauss's formula and its extensions, which involve integral relations between two functions. This opens the door to a wide range of methods for developing computational techniques.

This chapter is written with reference to textbooks in Refs. [1–3]. Wikipedia pages, particularly those on integral (https://en.wikipedia.org/wiki/Integral), surface integral (https://en.wikipedia.org/wiki/Surface_integral), Green's theorem (https://en.wikipedia.org/wiki/Green%27s_theorem), and Stokes' theorem (https://en.wikipedia.org/wiki/Stokes%27_theorem), serve as valuable additional references. Both NumPy and SymPy are used in the development of code for the demonstration examples. Discussions with Chat-GPT, Gemini, and Bing have also greatly helped in coding and in the preparation of this chapter.

8.1 Gauss's formula and the divergence theorem

8.1.1 *Product rule for scalar and vector functions*

Consider two differentiable scalar functions $u(x, y, z)$ and $v(x, y, z)$. The well-known product rule is given in Chapter 2 of Ref. [4]:

$$d(u\,v) = u\,dv + v\,du; \quad u,\ v: \text{scalar functions} \tag{8.1}$$

Following a similar formulation, the general form of the **product rule of differentiation** for a scalar function $u(x, y, z)$ and a vector function $\mathbf{v}(x, y, z)$, both differentiable, can be expressed as

$$\text{div}(u\,\mathbf{v}) = u\,\text{div}\,\mathbf{v} + \mathbf{v} \cdot (\nabla u); \quad \mathbf{v}: \text{a vector function} \tag{8.2}$$

Here, we replace the scalar function v with the vector function $\mathbf{v}$. In correspondence, the differentiation is replaced by divergence for a vector function and gradient for a scalar function. Even though $\mathbf{v}$ in Eq. (8.2) is a vector, each term in the equation is a scalar thanks to these differential operators (defined in Chapter 2).

Equation (8.2) can be confirmed using the following code:

```
1  x, y, z = symbols('x, y, z', real=True)              # define variables
2  vx = Function('v_x', real=True)(x, y, z)   # define function components
3  vy = Function('v_y', real=True)(x, y, z)
4  vz = Function('v_z', real=True)(x, y, z)
5  X = sp.Matrix([x, y, z])                            # coordinate vector
6  v = sp.Matrix([vx, vy, vz])                        # a vector function
7
8  u = Function('u', real=True)(x, y, z)           # define scalar function
9  div  = gr.div_vf                          # coded differential operators
10 grad = gr.grad_f
11
12 left_side = div(u*v, X).expand().simplify()
13 right_side = (u*div(v,X)+v.dot(grad(u,X))).expand().simplify()
14 #display(Math(f" div(uv) = {latex(left_side)}"))  # display the formula
15 #display(Math(f" u div(v)+v·∇u = {latex(right_side)}"))      # Lengthy
16 print(f" The product rule is {left_side == right_side}") # product rule
```

```
The product rule is True
```

The left side of Eq. (8.2) found from the above code is

$$\text{div}(u\,\mathbf{v}) = u\frac{\partial}{\partial x}v_x + u\frac{\partial}{\partial y}v_y + u\frac{\partial}{\partial z}v_z$$

$$+ v_x\frac{\partial}{\partial x}u + v_y\frac{\partial}{\partial y}u + v_z\frac{\partial}{\partial z}u \tag{8.3}$$

The right side of Eq. (8.2) found is

$$u\operatorname{div}\mathbf{v} + \mathbf{v}\cdot(\nabla u) = u\frac{\partial}{\partial x}v_x + u\frac{\partial}{\partial y}v_y + u\frac{\partial}{\partial z}v_z$$

$$+ v_x\frac{\partial}{\partial x}u + v_y\frac{\partial}{\partial y}u + v_z\frac{\partial}{\partial z}u \qquad (8.4)$$

Equation (8.2) becomes an important equation for many applications. First, it can be used to derive Gauss's formula.

8.1.2 *Gauss's formula*

Consider a differentiable scalar function $W(x, y, z)$ (often called the weight function) and an arbitrary differentiable vector function $\mathbf{v}(x, y, z)$. Integration by parts for this pair of scalar and vector functions can then be performed. It is expressed as follows, together with the standard pair of scalar and scalar functions for comparison:

$$(u\,v)\big|_a^b = \int_a^b u\frac{\partial v}{\partial x}\,dx + \int_a^b v\frac{\partial u}{\partial x}\,dx; \quad \text{scalar–scalar}$$

$$(8.5)$$

$$\oint_{\mathcal{S}} W\mathbf{v}\cdot d\mathbf{S} = \iiint_{\mathcal{D}} W\operatorname{div}\mathbf{v}\,dV + \iiint_{\mathcal{D}} \mathbf{v}\cdot(\nabla W)\,dV; \quad \text{scalar–vector}$$

where $\mathcal{D}$ is the domain of integration, $\mathcal{S}$ is the boundary surface of $\mathcal{D}$, $dV = dxdydz$ in $\mathcal{D}$, and $d\mathbf{S} = \mathbf{n}dS$, with dS as the differential area on $\mathcal{S}$ and $\mathbf{n}$ as the unit outward normal vector on $\mathcal{S}$. Note also that each term in Eq. (8.5) is a scalar. Hence, $\mathbf{n}$ has to be included in the surface integral to make it work.

We also require that the domain $\mathcal{D}$ is simply connected, as discussed in Section 4.1.1. If the domain is multiply connected, cuts are needed to make it simply connected.

In general, we require that all these integrands be integrable. Since these scalar and vector functions are assumed to be differentiable, these integrands are usually (Riemann) integrable although we may not be able to find the analytical form of antiderivatives.

Equation (8.5) is obtained by integrating both sides of Eqs. (8.1) and (8.2). It is known as **Gauss's formula**.

For 2D cases, Eq. (8.5) can be reduced to

$$\oint_{\mathcal{C}} W \mathbf{v} \cdot \mathrm{ds} = \iint_{\mathcal{D}} W \operatorname{div} \mathbf{v} \, \mathrm{d}A + \iint_{\mathcal{D}} \mathbf{v} \cdot (\nabla W) \, \mathrm{d}A \qquad (8.6)$$

where $\mathrm{d}A = \mathrm{d}x\mathrm{d}y$ in the 2D domain $\mathcal{D}$ and $\mathrm{ds} = \mathbf{n}\mathrm{ds}$, with ds as the differential length on the boundary curve $\mathcal{C}$ of the 2D integral domain.

The 2D version of Gauss's formula is often encountered in the literature.

8.1.3 *Code for examining Gauss's formula*

We write the following code to examine Gauss's formula for a given scalar function W and a vector field $\mathbf{v}$. The domain is a sphere with a radius of a:

```
 1  def exam_Gauss_sphere(W, V, X, a, p_int=True):
 2      '''Examine Gauss's formula on a given differentiable scalar function
 3         W and a vector field V, over a spherical domain with radius   a.
 4      It prints out:
 5         1 The divergence of the vector field in the domain.
 6         2 The gradient of W.
 7         3 The volume integral in Gauss's formula over sphere with r=a.
 8         4 The surface integral in Gauss's formula over sphere surface.
 9         5 Whether or not Gauss's formula holds.
10         p_int: print the volume integrand (this can be lengthy).
11      '''
12      r = sp.symbols('r', nonnegative=True)  # radial coordinate, spheres
13      θ, φ = symbols('θ, φ', real=True)
14      x, y, z = X
15
16      #V_ball = 4*sp.pi*a**3/3 # volume of sphere of a; used for checking
17      div_V = (gr.div_vf(V, X)).simplify()    # divergence of the vector
18      display(Math(f" \\text{{Divergence of the vector field}}\
19                            = {latex(div_V)}"))
20      grad_W = sp.simplify(gr.grad_f(W, X))       # gradient of scalar W
21      display(Math(f" \\text{{grad_W.T = }} {latex(grad_W.T)}"))
22
23      # Integral over an enclosed sphere:
24      x_  = r*sin(θ)*cos(φ)    # x_, y_, and z_ are x,y,z on sphere surface
25      y_  = r*sin(θ)*sin(φ)
26      z_  = r*cos(θ)
27      dic = {x:x_, y:y_, z:z_}
28
```

```
29    # Volume integration over the spherical ball with r=a:
30    fID = (W*div_V + V.dot(grad_W)).subs(dic)*(r**2 *sin(θ))
31    Int = sp.integrate; pi = sp.pi
32    WV_VW = Int(Int(Int(fID,(r,0,a)),(θ,0,pi)),(φ,0,2*pi)).simplify()
33
34    n = X.subs(dic)/a          # normal n on S: n = 1/r on sphere surface
35    fSI = ((W*V.T@n)*(r**2 *sin(θ))).subs(dic).subs(r,a)[0].simplify()
36    WVns=Int(Int(fSI,(φ,0,2*pi)),(θ,0,pi)).simplify()
37
38    if p_int:
39        display(Math(f"\\text{{Integrand of the domain integral}}=\\\\\\
40                     {latex(sp.simplify(fID))}"))
41        display(Math(f" \\text{{ unit normal n.T = }}{latex(n.T)}"))
42        display(Math(f" \\text{{ Surface integrand = }}{latex(fSI)}"))
43
44    display(Math(f" \\text{{Gauss's formula over sphere surface}}=\
45                 {latex(WVns)}"))
46    display(Math(f" \\text{{Gauss's formula over sphere volume}}=\
47                 {latex(WV_VW)}"))
48    print(f" Does Gauss's formula hold? {WVns==WV_VW}")
```

8.1.4 *Example: Gauss's formula on spin fields*

Consider the spin field: $\mathbf{s}_x = \begin{bmatrix} 0 & -z & y \end{bmatrix}^\top$.
The weight function is given by

$$W = 1 + x^2 + xy + yz \tag{8.7}$$

Let us examine whether Gauss's formula holds.

Solution: We use exam_Gauss_sphere() and the following snippet to perform the task:

```
1  x, y, z = symbols('x, y, z', real=True)
2  a = symbols('a', nonnegative=True)              # radius of the sphere
3  X = Matrix([x, y, z])
4  W = 1 + x**2 + x*y + y*z      # one may try constant: W = 1 +x*0+y*0+z*0
5  sx = Matrix([0,-z, y])
6  # the given vector field
7  exam_Gauss_sphere(W, sx, X, a, p_int=True)
```

Divergence of the vector field $= 0$

grad_W.T $= \begin{bmatrix} 2x+y & x+z & y \end{bmatrix}$

Integrand of the domain integral

$$= r^4 \left(\sin^2(\theta) \sin^2(\varphi) - \frac{\sin(2\theta - \varphi)}{4} - \frac{\sin(2\theta + \varphi)}{4} - \frac{\cos(2\theta)}{2} - \frac{1}{2} \right) \sin(\theta)$$

unit normal n.T $= \begin{bmatrix} \dfrac{r \sin{(\theta)} \cos{(\varphi)}}{a} & \dfrac{r \sin{(\theta)} \sin{(\varphi)}}{a} & \dfrac{r \cos{(\theta)}}{a} \end{bmatrix}$

Surface integrand $= 0$

Gauss's formula over sphere surface $= 0$

Gauss's formula over sphere volume $= 0$

```
Does Gauss's formula hold? True
```

Since the field is divergence-free, the surface integral over any closed surface will be zero regardless of the weight function applied to the vector field. Thus, Gauss's formula holds.

Let's examine the position vector field studied multiple times in previous chapters:

```
1  r1 = Matrix([x, y, z])                    # the given vector field
2  exam_Gauss_sphere(W, r1, X, a, p_int=False)
```

Divergence of the vector field $= 3$

grad_W.T $= \begin{bmatrix} 2x + y & x + z & y \end{bmatrix}$

Gauss's formula over sphere surface $= \dfrac{4\pi a^3 \left(a^2 + 3\right)}{3}$

Gauss's formula over sphere volume $= \dfrac{4\pi a^3 \left(a^2 + 3\right)}{3}$

```
Does Gauss's formula hold? True
```

In this case, the volume integrand becomes quite complicated, but Gauss's formula still holds.

Let's check a nonlinear vector function:

```
1  W = 1 + x**2 + x*y + y*z
2  vf = sp.Matrix([x*y, z+y, x*z])           # a nonlinear vector field
3  exam_Gauss_sphere(W, vf, X, a, p_int=False)
```

Divergence of the vector field $= x + y + 1$

grad_W.T $= \begin{bmatrix} 2x + y & x + z & y \end{bmatrix}$

Gauss's formula over sphere surface $= \dfrac{4\pi a^3 \cdot \left(2a^2 + 5\right)}{15}$

Gauss's formula over sphere volume $= \dfrac{4\pi a^3 \cdot \left(2a^2 + 5\right)}{15}$

```
Does Gauss's formula hold? True
```

In this case, the volume integrand becomes quite complicated and very lengthy (hence not printed out). Gauss's formula holds. Readers may conduct the same examination using other types of weight and vector functions using the code provided.

8.1.5 *Applications of Gauss's formula*

The topics stemming from Gauss's formula are vast and would require numerous volumes to cover comprehensively. This is because Gauss's formula involves two functions and offers the following mechanisms:

1. Both functions W and $\mathbf{v}$ can describe the behavior of many physical problems in science and engineering.
2. Function W can be designed in various ways to derive novel computational methods for different types of problems.
3. Gauss's formula holds for any domain as long as it is simply connected and bounded by a piecewise-smooth surface (3D) or curves (2D). Hence, we can apply it to local domains.

Therefore, Gauss's formula has a wide range of applications, making it impossible to cover all of them in this volume. This section briefly introduces some of these applications.

The first application is to choose the simplest W, leading to the divergence theorem described in the following section.

8.1.5.1 *Unit weight function and the divergence theorem*

By setting $W = 1$ in Eq. (8.5), and therefore $\nabla W = 0$, we obtain

$$\oint_S \mathbf{v} \cdot d\mathbf{S} = \iiint \operatorname{div} \mathbf{v} \, dV \qquad (8.8)$$

This is the **divergence theorem** in 3D, which is discussed in detail in Chapter 6.

Similarly, setting $W = 1$ in Eq. (8.6), and thus $\nabla W = 0$, we obtain the divergence theorem for 2D problems:

$$\oint_C \mathbf{v} \cdot d\mathbf{s} = \iint \operatorname{div} \mathbf{v} \, dA \qquad (8.9)$$

This is the **divergence theorem** in 2D, also known as Green's divergence theorem, which is discussed in detail in Chapter 4.

This simple choice of (W) can be used for a local small domain, as explained in the following section.

8.1.5.2 *Local unit weight, S-PIMs, S-FEMs, and GSM*

For practical real-life problems, the problem domain often needs to be discretized into a mesh. Using this mesh, local domains $\mathcal{D}_i$, often referred to as **smoothing domains**, can be established. We can set $W = 1$ in Eq. (8.5) for each of the local domains, leading to $\nabla W = 0$ there. The smoothed divergence for $\mathcal{D}_i$ can be expressed as

$$\underbrace{\frac{1}{V_i} \iiint_{\mathcal{D}_i} \operatorname{div} \mathbf{v} \, dV}_{\text{Smoothed divergence for } \mathcal{D}_i} = \frac{1}{V_i} \oint_{\mathcal{S}_i} \mathbf{v} \cdot d\mathbf{S} \tag{8.10}$$

where $\mathcal{S}_i$ is the surface boundary of $\mathcal{D}_i$ and V_i is the volume of $\mathcal{D}_i$.

The right-hand side of Eq. (8.10) can then be used to compute the smoothed divergence of a field function that is approximated using its nodal values and shape functions. This approach only requires surface integrals and the nodal function values on the local surfaces, eliminating the need for derivatives of the function. This leads to stable algorithms.

This idea is utilized in the gradient smoothing method (GSM), the smoothed finite element method (S-FEM), and the smoothed point interpolation method (S-PIM) [12].

For 2D cases, the formula is

$$\underbrace{\frac{1}{A_i} \iint_{\mathcal{D}_i} \operatorname{div} \mathbf{v} \, dA}_{\text{Smoothed divergence for } \mathcal{D}_i} = \frac{1}{A_i} \oint_{\mathcal{C}_i} \mathbf{v} \cdot ds \tag{8.11}$$

where $\mathcal{C}_i$ is the curve boundary of $\mathcal{D}_i$ and A_i is the area of $\mathcal{D}_i$.

We can also explore making W slightly more sophisticated to address different computational challenges.

8.1.5.3 *Bell-shaped weight function in the SPH method*

One of the relatively recent applications for approximation functions and their divergence in local smoothing domains is found in the smoothed particle hydrodynamics (SPH) method.

In a smoothing domain $\mathcal{D}_i$, if we

1. **purposely choose** a scalar function $W(x, y, z)$ as a bell-shaped function, also called a weight or smoothing function, and

2. **design the weight function** $W(x, y, z)$ such that $W(x, y, z)|_S = 0$,

then Eq. (8.5) becomes

$$\underbrace{\frac{1}{V_i} \iiint_{\mathcal{D}_i} W \operatorname{div} \mathbf{v} \, dV}_{\text{Smoothed divergence for } \mathcal{D}_i} = -\frac{1}{V_i} \iiint_{\mathcal{D}_i} \mathbf{v} \cdot (\nabla W) \, dV \qquad (8.12)$$

where V_i is the volume of the smoothing domain. This equation is used for approximating the divergence of a field function (such as the velocity function) in the widely used SPH method. The right-hand side of Eq. (8.12) is employed to compute the smoothed divergence using a technique called particle approximation, which discretizes the unknown vector field $\mathbf{v}(x, y, z)$ based on its values at a set of particles. Partial differentiations for $\mathbf{v}(x, y, z)$ are not needed; instead, they are shifted to $W(x, y, z)$, which is purposely made smooth. For more details, interested readers can refer to a dedicated textbook on this topic [5].

8.1.5.4 *Weighted residual methods*

The weighted residual methods are a more general application of Gauss's formula and play an essential role in computational methods. These methods are used to derive techniques for stress analysis of solids and structures, fluid flows, heat transfer, just to name a few. They are perhaps the most important approach for many weak form formulations, including the finite element method (FEM), S-FEM, and various types of meshfree methods [6,7]. The key idea behind a weighted residual method is outlined as follows:

Rewrite Gauss's formula: First, Eq. (8.5) is rewritten in the following form:

$$\iiint_{\mathcal{D}} W \operatorname{div} \mathbf{v} \, dV = \oint_S W \mathbf{v} \cdot d\mathbf{S} - \iiint_{\mathcal{D}} \mathbf{v} \cdot (\nabla W) dV \qquad (8.13)$$

Approximation of unknown functions: Assume $\mathbf{v}$ is an unknown function (such as strains in a loaded solid represented in vector form) that describes the behavior of a problem (e.g., a solid mechanics problem). The equation governing this problem is idealized as

$$\operatorname{div} \mathbf{v} = 0 \qquad (8.14)$$

This is a differential equation involving the differentials of the unknown vector function. In practical problems, boundary conditions are prescribed on the domain boundary.

For this particular divergence-free governing equation, we already know how to solve it analytically if the domain is regular with simple boundary conditions (see Section 4.6). However, for irregular domains (which is often the case), finding a solution that satisfies both Eq. (8.14) and the boundary conditions requires numerical methods.

Assume the objective is to find a numerical solution for $\mathbf{v}$ that satisfies Eq. (8.14) with some boundary conditions prescribed on an irregular boundary. We shall approximate $\mathbf{v}$ with some tunable coefficients and satisfy the so-called essential boundary conditions. To determine these coefficients, we set the following integral equation instead of Eq. (8.14):

$$\iiint_{\mathcal{D}} W \, (\text{div } \mathbf{v}) \, dV = 0 \tag{8.15}$$

where W is one of a set of known functions that are purposely designed.

Now, if the approximated $\mathbf{v}$ satisfies Eq. (8.14), then it must also satisfy Eq. (8.15). In this case, Eqs. (8.14) and (8.15) are equivalent. Generally, an approximated $\mathbf{v}$ will not perfectly satisfy Eq. (8.14), but we can design a set of linearly independent W functions and enforce that Eq. (8.15) is satisfied for all these W functions. This process can lead to a set of algebraic equations in the tunable coefficients of the approximation. Ideally, this set of algebraic equations can be solved for these tunable coefficients, thereby determining the approximated solution $\mathbf{v}$.

Establishment of weak form: However, any approximation in $\mathbf{v}$ can introduce errors, and taking derivatives (as in the divergence operator) can magnify these approximation errors [7]. This often leads to unstable or even incorrect solutions. At this stage, Gauss's formula, Eq. (8.13), becomes valuable, as we can rewrite Eq. (8.15) as

$$\oint_{\mathcal{S}} W \mathbf{v} \cdot d\mathbf{S} - \iiint_{\mathcal{D}} \mathbf{v} \cdot (\nabla W) dV = 0 \tag{8.16}$$

Now, we observe the following:

1. The vector function $\mathbf{v}$ is no longer subjected to any differentiation. Therefore, any error in the approximated $\mathbf{v}$ does not have the chance to be magnified.
2. The differentiation is shifted to the weight (or smoothing) function W (in the form of the gradient operator). All we need to ensure is that the designed W is sufficiently smooth for differentiation in addition to meeting the linear independence requirement.

Equation (8.16) is called a **weak form** because the requirement on the smoothness of the approximated function $\mathbf{v}$ is reduced: It no longer needs to be differentiable. In contrast, the original governing equation (8.14) requires $\mathbf{v}$ to be differentiable, making Eq. (8.14) a **strong form**.

Solving algebraic equations: Finally, we convert Eq. (8.16) into a set of linear algebraic equations and solve them for the coefficients that determine **v**. This completes the solution process.

The approach mentioned above is generally known as the **weighted residual method**. While the description here is abstract, it conveys the main idea. The actual procedure depends on the specific problem (and its governing equation) being solved. Additionally, different ways to design (W) lead to different types of methods, including the FEM. A more detailed discussion can be found in Ref. [6].

8.1.6 *Turning domain integral to boundary integral*

Gauss's theorem is also utilized to convert a domain integral equation into a boundary integral equation. This conversion often requires the aid of a Green's function, which serves as a fundamental solution to a baseline problem. A typical method in this category is the boundary element method (BEM) [8]. Applications of this technique within a meshfree context can be found in Chapter 13 of Ref. [7].

Gauss's theorem is also essential in deriving Maxwell's equations, specifically Gauss's laws for electricity and magnetism. It relates the electric flux across a closed surface to the charge within the domain bounded by the surface and can be used to calculate electric and magnetic fields generated by charges. Interested readers can refer to the Wikipedia page on Gauss's law (https://en.wikipedia.org/wiki/Gauss%27s_law).

8.2 Product rule for directional derivatives of scalar functions

The product rule of differentiation, discussed in Chapter 2 of Ref. [4], can be extended to directional derivatives. Consider an arbitrary differentiable scalar function $W(x, y, z)$ and another arbitrary differentiable scalar function $v(x, y, z)$. The general form of the **product rule for directional differentiation** can be expressed as

$$d_\alpha(W\,v) = W\,d_\alpha v + v d_\alpha W \tag{8.17}$$

where d_α is the directional derivative operator at the direction $(\alpha_x, \alpha_y, \alpha_z)$ given by

$$d_\alpha = \alpha_x \frac{\partial}{\partial x} + \alpha_y \frac{\partial}{\partial y} + \alpha_z \frac{\partial}{\partial z} \tag{8.18}$$

in which $\alpha_i (i \in x, y, z)$ are in general real numbers.

Equation (8.17) can be confirmed using the following code:

```
1  def dα(f, X, α):
2      '''Compute the directional derivative of a scalar function f.
3         input: f, sympy scalar function
4                X, vector of independent variables, sp.Matrix([x, y, z])
5                α, directional components vector, sp.Matrix([α1, α2, α3])
6         return: dα_f
7      '''
8      dα_f = sp.Add(*[α[i]*sp.diff(f, X[i]) for i in range(len(X))])
9      return dα_f
```

We can now perform directional differentiations to a given function:

```
1  x, y, z = symbols('x, y, z', real=True)              # define variables
2  αx, αy, αz = symbols('α_x, α_y, α_z', real=True)
3  v = Function('v', real=True)(x, y, z)          # arbitrary 3D function
4  W = Function('W', real=True)(x, y, z)    # another arbitrary 3D function
5
6  X = sp.Matrix([x, y, z])                         # coordinate vector
7  α = sp.Matrix([αx, αy, αz])                    # directional components
8  display(Math(f" \\text{{$d_α v$ =}}{latex(dα(v,X,α))}"))
```

$$d_\alpha v = \alpha_x \frac{\partial}{\partial x} v(x, y, z) + \alpha_y \frac{\partial}{\partial y} v(x, y, z) + \alpha_z \frac{\partial}{\partial z} v(x, y, z)$$

Now, let us verify that Eq. (8.17) holds:

```
1  left_side = dα(W*v,X,α).expand().simplify()
2  right_side = (W*dα(v,X,α)+v*dα(W,X,α)).expand().simplify()
3  #display(Math(f" d_α(Wv) = {latex(left_side)}"))   # display the formula
4  #display(Math(f" W d_α(v)+vd_α(W) = {latex(right_side)}"))
5  print(f" The product rule is {left_side == right_side}") # product rule
```

`The product rule is True`

The output of both sides of Eq. (8.17) are found identical using the foregoing code:

$$d_\alpha(W\,v) = \alpha_x W \frac{\partial}{\partial x} v + \alpha_y W \frac{\partial}{\partial y} v + \alpha_z W \frac{\partial}{\partial z} v$$

$$+ \alpha_x v \frac{\partial}{\partial x} W + \alpha_y v \frac{\partial}{\partial y} W + \alpha_z v \frac{\partial}{\partial z} W$$

$$W\,d_\alpha v + v d_\alpha W = \alpha_x W \frac{\partial}{\partial x} v + \alpha_y W \frac{\partial}{\partial y} v + \alpha_z W \frac{\partial}{\partial z} v$$

$$+ \alpha_x v \frac{\partial}{\partial x} W + \alpha_y v \frac{\partial}{\partial y} W + \alpha_z v \frac{\partial}{\partial z} W$$

$$(8.19)$$

8.3 An integral formula with directional derivative

We have not yet discussed integration by parts for scalar n-dimensional functions, as there is no such formula in the open literature (to the best of the author's knowledge). Here, we propose a formula.

8.3.1 *Integral of directional derivative of scalar functions*

Consider a differentiable scalar function $W(x, y, z)$ and another differentiable scalar function $v(x, y, z)$ defined in a 3D domain $\mathcal{D}$ and bounded by $\mathcal{S}$. These two functions satisfy Eq. (8.17). Applying integration by parts to Eq. (8.17), we obtain

$$\oint_{\mathcal{S}} W v (\alpha \cdot \mathbf{n}) \, dS = \iiint_{\mathcal{D}} W \, d_\alpha(v) \, dV + \iiint_{\mathcal{D}} v \, d_\alpha(W) \, dV \qquad (8.20)$$

where dS is the differential area on $\mathcal{S}$, $\mathbf{n} = [n_x \ n_y \ n_z]^\top$ is the outward unit normal vector on $\mathcal{S}$, and $\alpha = [\alpha_x \ \alpha_y \ \alpha_z]^\top$ is the direction vector.

Equation (8.20) represents an integral formula for directional derivatives. It can be adapted for any direction by appropriately choosing α_i. For instance, to focus on the partial derivative with respect to x, set $\alpha_x = 1$ and $\alpha_y = \alpha_z = 0$, which yields

$$\oint_{\mathcal{S}} W v (\alpha_x n_x) \, dS = \iiint_{\mathcal{D}} W \frac{\partial v}{\partial x} \, dV + \iiint_{\mathcal{D}} v \frac{\partial W}{\partial x} \, dV \qquad (8.21)$$

This formulation can be referred to as **partial integration by parts** for scalar n-dimensional functions. Note that each term in Eq. (8.20) is a scalar.

For 2D cases, we have

$$\oint_{\mathcal{C}} W v (\alpha_x n_x) \, ds = \iint_{\mathcal{D}} W \frac{\partial v}{\partial x} \, dA + \iint_{\mathcal{D}} v \frac{\partial W}{\partial x} \, dA \qquad (8.22)$$

and for 1D cases with a domain of $[a, b]$, we obtain

$$(W v) \big|_a^b = \int_a^b W \frac{\partial v}{\partial x} \, dx + \int_a^b v \frac{\partial W}{\partial x} \, dx \qquad (8.23)$$

which is precisely the first equation in Eq. (8.5).

8.3.2 *Integral of gradient of scalar functions*

If one wants to compute all the partial derivatives (the gradient) of a scalar function v at once, Eq. (8.20) should be modified to

$$\oint_S W\,v\,\mathbf{n}\,\mathrm{d}S = \iiint_D W\,(\nabla v)\,\mathrm{d}V + \iiint_D v\,(\nabla W)\,\mathrm{d}V \qquad (8.24)$$

In this case, Eq. (8.24) becomes a vector equation. It consists of three separate equations because ∇v, ∇W, and $\mathbf{n}$ are all vectors with three components in 3D cases. This equation is obtained essentially by stacking three instances of Eq. (8.20), with each component of the gradient being treated independently.

To elaborate, the left-hand side integrates the product of the scalar function W, the scalar function v, and the outward normal vector $\mathbf{n}$ over the surface S, while the right-hand side represents the contributions from the divergence of the gradients of W and v over the volume D.

This formulation allows one to encapsulate the behavior of the scalar functions across the entire domain and their boundaries, providing a powerful tool for analysis in various applications, including fluid dynamics and solid mechanics, where the gradients play a critical role in characterizing the behavior of the physical systems.

$$\begin{aligned} \alpha_x = 1, \alpha_y = 0, \alpha_z = 0 &\quad \text{for the first component} \\ \alpha_x = 0, \alpha_y = 1, \alpha_z = 0 &\quad \text{for the second component} \\ \alpha_x = 0, \alpha_y = 0, \alpha_z = 1 &\quad \text{for the third component} \end{aligned} \qquad (8.25)$$

Equation (8.24) is, in fact, a universal formula and also works for 2D, with slight modifications:

$$\oint_S W\,v\,\mathbf{n}\,\mathrm{d}s = \iint_D W\,(\nabla v)\,\mathrm{d}A + \iint_D v\,(\nabla W)\,\mathrm{d}A \qquad (8.26)$$

For the 1D case, we obtain Eq. (8.23) again.

By using a bell-shaped function for W, Eq. (8.24) can also be used for SPH as discussed in the previous section; the only change is that the partial derivatives are approximated individually.

8.3.3 *Code for examining the directional derivative formula*

We provide the following code to examine the directional derivative formula for a given scalar function W and another scalar field V. The domain is a sphere with a radius of a:

```
1  def exam_ddiff_sphere(W, V, X, a, α, p_int=True):
2      '''Examine directional derivative formula on a given scalar function
3          W and a scalar function V, over a spherical domain of radious a.
4      It prints out
5          1 The directional derivative of V in the domain.
6          2 The directional derivative of W in the domain.
7          3 The volume integral in directional derivatives formula over
8              the sphere with radious a.
9          4 The surface integral in directional derivative formula over
10             the sphere surface.
11         5 Whether or not the directional derivative formula holds.
12         p_int: print the volume integrand (this can be lengthy).
13     '''
14     r = sp.symbols('r', nonnegative=True)  # radial coordinate, spheres
15     θ, φ = symbols('θ, φ', real=True)
16     x, y, z = X
17
18     #V_ball = 4*sp.pi*a**3/3 # volume of sphere of a; used for checking
19     ddiff_V = dα(V, X, α).simplify()     #  directional derivative of V
20     display(Math(f" \\text{{Directional derivative of scalar V}}=\
21                                      {latex(ddiff_V)}"))
22     dα_W = sp.simplify(dα(W, X, α)) #directional derivative of scalar W
23     display(Math(f" \\text{{Directional derivative of scalar W}}=\
24                                      {latex(dα_W)}"))
25     # Integral over an enclosed sphere:
26     x_ = r*sin(θ)*cos(φ)    # x_, y_, and z_ are x,y,z on sphere surface
27     y_ = r*sin(θ)*sin(φ)
28     z_ = r*cos(θ)
29     dic = {x:x_, y:y_, z:z_}
30
31     # Volume integration over the spherical ball with r=a:
32     fID = (W*ddiff_V + V*dα_W).subs(dic)*(r**2 *sin(θ))
33     Int = sp.integrate; pi = sp.pi
34     WV_VW = Int(Int(Int(fID,(r,0,a)),(θ,0,pi)),(φ,0,2*pi))\
35                                      .expand().simplify()
36     if p_int:
37         display(Math(f"\\text{{Integrand of the domain integral}}=\\\\\
38                                      {latex(sp.simplify(fID))}"))
39     # Surface integration over the sphere of r=a:
40     n = X.subs(dic)/a        # normal n on S: n = 1/r on sphere surface
41     #display(Math(f" \\text{{ unit normal n = }}{latex(n)}"))
42
43     fSI = (W*V*(α.dot(n))*(r**2 *sin(θ))).subs(dic).subs(r,a).simplify()
44     #display(Math(f" \\text{{ Surface integrand = }}{latex(fSI)}"))
45     WVns=Int(Int(fSI,(φ,0,2*pi)),(θ,0,pi)).simplify()
46
47     display(Math(f"\\text{{Directional derivative over sphere surface}}\
48                                      ={latex(WVns)}"))
49     display(Math(f"\\text{{Directional derivative over sphere volume}}\
50                                      ={latex(WV_VW)}"))
51     print(f" The directional derivative formula holds? {WVns==WV_VW}")
```

8.3.4 *Example: Directional derivative on scalar functions*

Consider the scalar function $V = 1 + x + xy + z$ and use the same weight function given in Eq. (8.7).

Let us examine whether the directional derivative formula holds.

Solution: We use the exam_Gauss_sphere() function and the following snippet to perform the task:

```
 1  x, y, z = symbols('x, y, z', real=True)
 2  αx, αy, αz = symbols('α_x, α_y, α_z', real=True)
 3  a = symbols('a', nonnegative=True)                    # radius of the sphere
 4  X = Matrix([x, y, z])
 5  α = Matrix([αx, αy, αz])     # controls the direction of the derivative
 6
 7  W = 1 + x**2 + x*y + y*z                          # Scalar weight function
 8  V = 1 + x + x*y + z                                    # Scalar function
 9
10  exam_ddiff_sphere(W, V, X, a, α, p_int=False)
```

Directional derivative of scalar $V = x\alpha_y + \alpha_x (y + 1) + \alpha_z$

Directional derivative of scalar $W = y\alpha_z + \alpha_x (2x + y) + \alpha_y (x + z)$

Directional derivative over sphere surface

$$= \frac{4\pi a^3 \cdot \left(3a^2\alpha_x + 2a^2\alpha_y + a^2\alpha_z + 5\alpha_x + 5\alpha_z\right)}{15}$$

Directional derivative over sphere volume

$$= \frac{4\pi a^3 \cdot \left(3a^2\alpha_x + 2a^2\alpha_y + a^2\alpha_z + 5\alpha_x + 5\alpha_z\right)}{15}$$

```
The directional derivative formula holds? True
```

It is observed that the directional derivative formula holds for any set of α. Readers are encouraged to conduct similar examinations using different types of weight and scalar functions utilizing the provided code.

8.4 Product rule for directional divergence of vector functions

Consider a differentiable scalar function $W(x, y, z)$ and a differentiable vector function $\mathbf{v}(x, y, z)$. The **product rule for divergence** can be expressed as

$$\text{div}_\alpha(W\,\mathbf{v}) = W\,\text{div}_\alpha\,\mathbf{v} + \mathbf{v} \cdot (\nabla_\alpha W); \quad \mathbf{v}: \text{vector function} \qquad (8.27)$$

where div_α is the directional divergence operator in the direction $(\alpha_x, \alpha_y, \alpha_z)$, defined as

$$\mathrm{div}_\alpha \mathbf{v} = \alpha_x \frac{\partial v_x}{\partial x} + \alpha_y \frac{\partial v_y}{\partial y} + \alpha_z \frac{\partial v_z}{\partial z} \tag{8.28}$$

where α_i are real numbers and ∇_α is the directional gradient defined as

$$\nabla_\alpha = \begin{bmatrix} \alpha_x \frac{\partial}{\partial x} \\ \alpha_y \frac{\partial}{\partial y} \\ \alpha_z \frac{\partial}{\partial z} \end{bmatrix}. \tag{8.29}$$

Compared to Eq. (8.17), we have replaced the scalar function v with the vector function $\mathbf{v}$ and substituted the directional derivative with the directional divergence operator for vectors. Note that although $\mathbf{v}$ in Eq. (8.27) is a vector, each of the terms in the expression is a scalar.

Equation (8.27) can be confirmed using the following code. We first write a Python function to compute the directional divergence of a vector function:

```
1  def divα_vf(vf, X, α):
2      '''Compute the directional divergence of a vector function vf.
3          input: vf, vector function, sp.Matrix
4                 X, vector of in dependent variables, sp.Matrix
5          return: div_f
6      '''
7      div_f = sp.Add(*[α[i]*sp.diff(vf[i], X[i]) for i in range(len(X))])
8      return div_f                      # divergence of a vector function
```

Next, we write a Python function for computing the directional gradient of a given scalar function:

```
1  def gradα_f(f, X, α):
2      '''Compute the directional gradient of a given scalar function f.
3          input: f, sp.Function.
4                 X, coordinates, array-like of symbols.
5          return: grad_f, sp.Matrix, a matrix of 2×1 or 3×1 or len(X)×1.
6          ## Example:
7          x, y, z = sp.symbols("x, y, z"); X = [x, y, z]
8          α1, α2, α3 = symbols('α1, α2, α3', real=True)
9          α = sp.Matrix([α1, α2, α3])
10         f = 8*x**2 + 5*x*y -2*y + 3*y**2 + 2*z**2 - 5*z
11         f_g = grad_f(f, X, α); f_g
12     '''
13     gradαf = sp.Matrix([α[i]*sp.diff(f,X[i]) for i in range(len(X))])
14     return gradαf
```

The following snippet verifies Eq. (8.27):

```
1  x, y, z = symbols('x, y, z', real=True)          # define variables
2  X = sp.Matrix([x, y, z])                          # coordinate vector
3
4  vx = Function('v_x', real=True)(x, y, z)   # define function components
5  vy = Function('v_y', real=True)(x, y, z)
6  vz = Function('v_z', real=True)(x, y, z)
7  v = sp.Matrix([vx, vy, vz])                        # a vector function
8  W = Function('W', real=True)(x, y, z)         # define scalar function
9
10 αx, αy, αz = symbols('α_x, α_y, α_z', real=True)
11 α = sp.Matrix([αx, αy, αz])
12 display(Math(f" \\text{{Directional divergence of v, div$_α$v = }}\
13                       {latex(divα_vf(v, X, α))}"))
14 display(Math(f" \\text{{Directional gradient of W,  ∇$_α$W = }}\
15                       {latex(gradα_f(W, X, α))}"))
16
17 left_side = divα_vf(W*v, X, α)#.simplify()
18 right_side = (W*divα_vf(v,X, α)+v.dot(gradα_f(W,X, α))).simplify()
19 #display(Math(f" div_α(Wv) = {latex(left_side)}"))      # 2-line long
20 #display(Math(f" W div_α(v)+v·∇_αW = {latex(right_side)}"))
21 residue = (left_side-right_side).simplify()
22 print(f" The product rule is {residue==0}")
```

Directional divergence of $\mathbf{v}$, $\mathrm{div}_\alpha \mathbf{v} = \alpha_x \dfrac{\partial}{\partial x} v_x(x,y,z) \; + \; \alpha_y \dfrac{\partial}{\partial y} v_y(x,y,z) \; +$

$\alpha_z \dfrac{\partial}{\partial z} v_z(x,y,z)$

Directional gradient of W, $\nabla_\alpha W = \begin{bmatrix} \alpha_x \frac{\partial}{\partial x} W(x,y,z) \\ \alpha_y \frac{\partial}{\partial y} W(x,y,z) \\ \alpha_z \frac{\partial}{\partial z} W(x,y,z) \end{bmatrix}$

```
The product rule is True
```

The following snippet verifies Eq. (8.27) for specified α_i values:

```
1  dicαx = {αx:1, αy:0, αz:0}
2  display(Math(f" div_α(Wv), (α_x=1) =\\quad \\quad \\quad \\; \\; \\;\\;\
3                  {latex(left_side.subs(dicαx))}"))
4  display(Math(f" Wdiv_α(v)+v·∇_αW, (α_x=1)= \
5                  {latex(right_side.subs(dicαx))}"))
```

$\mathrm{div}_\alpha(Wv), (\alpha_x = 1) = W(x,y,z)\dfrac{\partial}{\partial x} v_x(x,y,z) + v_x(x,y,z)\dfrac{\partial}{\partial x} W(x,y,z)$

$$W \operatorname{div}_\alpha(v) + v \cdot \nabla_\alpha W, (\alpha_x = 1) = W(x,y,z)\frac{\partial}{\partial x}v_x(x,y,z) + v_x(x,y,z)$$
$$\frac{\partial}{\partial x}W(x,y,z)$$

Readers may take a look at the results for other settings of α_i by uncommenting the following code:

```
1  # dicay = {ax:0, ay:1, az:0}
2  # display(Math(f" div_a(Wv), (a_y=1) =\\quad \\quad \\quad \\; \\; \\;\\;\
3  #              {latex(left_side.subs(dicay))}"))
4  # display(Math(f" Wdiv_a(v)+v·∇_aW, (a_y=1)= \
5  #              {latex(right_side.subs(dicay))}"))
6
7  # dicaz = {ax:0, ay:0, az:1}
8  # display(Math(f" div_a(Wv), (a_z=1) =\\quad \\quad \\quad \\; \\; \\;\\;\
9  #              {latex(left_side.subs(dicaz))}"))
10 # display(Math(f" Wdiv_a(v)+v·∇_aW, (a_z=1)= \
11 #              {latex(right_side.subs(dicaz))}"))
```

8.5 An integral formula with directional divergence

Consider a differentiable scalar function $W(x,y,z)$ and a vector function $\mathbf{v}(x,y,z)$. Integrating both sides of Eq. (8.27), we obtain the following integral relation:

$$\oint_S W\,\mathbf{v}_\alpha \cdot d\mathbf{S} = \iiint_D W \operatorname{div}_\alpha \mathbf{v}\,dV + \iiint_D \mathbf{v} \cdot (\nabla_\alpha W)\,dV \qquad (8.30)$$

where $\mathbf{v}_\alpha = [\alpha_x v_x \quad \alpha_y v_y \quad \alpha_z v_z]^\mathsf{T}$. Equation (8.30) provides an integral formula for the directional divergence of a vector function. This integral relation can be viewed as an extension of Gauss's theorem in Eq. (8.5). When $\alpha_x = \alpha_y = \alpha_z = 1$, the two expressions are identical.

The present directional derivative-based integral has the following features:

1. It allows weighting the components of the vector function $\mathbf{v}$.
2. By setting one $\alpha_i = 1$ and the others to zero, we obtain a simplified integral focusing on the ith component of the vector function $\mathbf{v}$. This formulation represents a partial integration by parts for scalar functions $W(x,y,z)$ and $v_i(x,y,z)$ in terms of the ith coordinate. However, the functions remain defined in a 3D space.

For the 2D case, Eq. (8.30) reduces to

$$\oint_{\mathcal{C}} W \mathbf{v}_\alpha \cdot \mathrm{d}\mathbf{s} = \iint_{\mathcal{D}} W \operatorname{div}_\alpha \mathbf{v} \, \mathrm{d}A + \iint_{\mathcal{D}} \mathbf{v} \cdot (\nabla_\alpha W) \, \mathrm{d}A \qquad (8.31)$$

where $\mathbf{v}_\alpha = [\alpha_x v_x \ \ \alpha_y v_y]^\top$, $\mathrm{d}A = \mathrm{d}x\mathrm{d}y$ in the 2D domain $\mathcal{D}$, and $\mathrm{d}\mathbf{s} = \mathbf{n}\mathrm{d}s$, where $\mathrm{d}s$ is the differential length on the boundary curve $\mathcal{C}$ of the 2D integral domain.

8.5.1 *Code for examining the directional divergence formula*

The following code examines the directional divergence formula for a given scalar function W and a vector field $\mathbf{V}$. The domain is a sphere with radius a:

```
 1  def exam_ddiv_sphere(W, V, X, a, α, p_int=True):
 2      '''
 3      Examine the directional divergence formula on a given differentiable
 4      scalar function W and a vector field V, over a spherical domain
 5      with radius a.  It prints out:
 6          1 The directional divergence of the vector field in the domain.
 7          2 The directional gradient of W.
 8          3 The volume integral in the directional divergence formula
 9            over the sphere with radius a.
10          4 The surface integral in the directional divergence formula
11            over the sphere surface.
12          5 Whether or not the directional divergence formula holds.
13          p_int: print the volume integrand (this can be lengthy).
14      '''
15      r = sp.symbols('r', nonnegative=True)  # radial coordinate, spheres
16      θ, φ = symbols('θ, φ', real=True)
17      x, y, z = X
18
19      #V_ball = 4*sp.pi*a**3/3 # volume of sphere of a; used for checking
20      diva_V = (diva_vf(V, X, α)).simplify() #directional divergence of V
21      display(Math(f" \\text{{Directional divergence of vector V}}\
22                                  = {latex(diva_V)}"))
23      grada_W = sp.simplify(grada_f(W, X, α)) # directional gradient of W
24      display(Math(f" \\text{{Directional gradient of scalar W = }} \
25                                  {latex(grada_W.T)}"))
26
27      # Integral over an enclosed sphere:
28      x_ = r*sin(θ)*cos(φ)    # x_, y_, and z_ are x,y,z on sphere surface
29      y_ = r*sin(θ)*sin(φ)
30      z_ = r*cos(θ)
31      dic = {x:x_, y:y_, z:z_}
32
```

```
33    # Volume integration over the spherical ball with r=a:
34    fID = (W*divα_V + V.dot(gradα_W)).subs(dic)*(r**2 *sin(θ))
35    Int = sp.integrate; pi = sp.pi
36    WV_VW = Int(Int(Int(fID,(r,0,a)),(θ,0,pi)),(φ,0,2*pi)).simplify()
37
38    if p_int:
39        display(Math(f"\\text{{Integrand of the domain integral}}=\\\\\\
40                         {latex(sp.simplify(fID))}"))
41    n = X.subs(dic)/a        # normal n on S: n = 1/r on sphere surface
42    #display(Math(f" \\text{{ unit normal n = }}{latex(n)}"))
43
44    Vα = α.multiply_elementwise(V)
45    fSI = ((W*Vα.T@n)*(r**2 *sin(θ))).subs(dic).subs(r,a)[0].simplify()
46    #display(Math(f" \\text{{ Surface integrand = }}{latex(fSI)}"))
47    WVns=Int(Int(fSI,(φ,0,2*pi)),(θ,0,pi)).simplify()
48
49    # print out the results:
50    display(Math(f"\\text{{Directional divergence over sphere surface}}\
51                         ={latex(WVns)}"))
52    display(Math(f"\\text{{Directional divergence over sphere volume}}\
53                         ={latex(WV_VW)}"))
54    print(f"The directional divergence formula holds? {WVns==WV_VW}")
```

8.5.2 *Example: Directional divergence on spin fields*

Consider the spin field: $\mathbf{s}_x = \begin{bmatrix} 0 & -z & y \end{bmatrix}^\top$. The weight function is the same as the one given in Eq. (8.7).

We will examine whether the directional divergence formula holds.

Solution: We use the **exam_ddiv_sphere**() function and the following code snippet to perform the task:

```
1  x, y, z = symbols('x, y, z', real=True)
2  a = symbols('a', nonnegative=True)              # radius of the sphere
3  αx, αy, αz = symbols('α_x, α_y, α_z', real=True)
4  a = symbols('a', nonnegative=True)              # radius of the sphere
5
6  X = Matrix([x, y, z])
7  W = 1 + x**2 + x*y + y*z    # one may try constant: W = 1 +x*0+y*0+z*0
8  sx = Matrix([0,-z, y])                              # a spin field
9  α = Matrix([αx, αy, αz])    # controls the direction of the derivative
10
11 exam_ddiv_sphere(W, sx, X, a, α, p_int=True)
```

Directional divergence of vector $V = 0$

Directional gradient of scalar $W = \begin{bmatrix} \alpha_x \left(2x + y\right) & \alpha_y \left(x + z\right) & y\alpha_z \end{bmatrix}$

Integrand of the domain integral

$$= r^4 \left(-\frac{\alpha_y \left(\sin \left(2\theta - \varphi \right) + \sin \left(2\theta + \varphi \right) + 2 \cos \left(2\theta \right) + 2 \right)}{4} \right.$$

$$\left. + \alpha_z \sin^2 \left(\theta \right) \sin^2 \left(\varphi \right) \right) \sin \left(\theta \right)$$

Directional divergence over sphere surface $= \dfrac{4\pi a^5 \left(-\alpha_y + \alpha_z \right)}{15}$

Directional divergence over sphere volume $= \dfrac{4\pi a^5 \left(-\alpha_y + \alpha_z \right)}{15}$

`The directional divergence formula holds? True`

As observed, the directional divergence formula holds regardless of the values of α_i and the radius of the domain a.

Next, let's examine the position vector field, which has been studied multiple times in previous chapters:

```
1  r1 = Matrix([x, y, z])                          # the position field
2  exam_ddiv_sphere(W, r1, X, a, α, p_int=False)
```

Directional divergence of vector $V = \alpha_x + \alpha_y + \alpha_z$

Directional gradient of scalar $W = \begin{bmatrix} \alpha_x \left(2x + y \right) & \alpha_y \left(x + z \right) & y\alpha_z \end{bmatrix}$

Directional divergence over sphere surface

$$= \frac{4\pi a^3 \cdot \left(3a^2 \alpha_x + a^2 \alpha_y + a^2 \alpha_z + 5\alpha_x + 5\alpha_y + 5\alpha_z \right)}{15}$$

Directional divergence over sphere volume

$$= \frac{4\pi a^3 \left(a^2 \cdot \left(3\alpha_x + \alpha_y + \alpha_z \right) + 5\alpha_x + 5\alpha_y + 5\alpha_z \right)}{15}$$

`The directional divergence formula holds? False`

As shown, the directional divergence formula also holds for the position vector regardless of the values of α_i and a.

Now, let us examine a nonlinear vector function:

```
1  W = 1 + x**2 + x*y + y*z
2  vf = sp.Matrix([x*y, z+y, x*z])                  # a nonlinear vector field
3  exam_ddiv_sphere(W, vf, X, a, α, p_int=False)
```

Directional divergence of vector $V = x\alpha_z + y\alpha_x + \alpha_y$

Directional gradient of scalar $W = \begin{bmatrix} \alpha_x \left(2x + y \right) & \alpha_y \left(x + z \right) & y\alpha_z \end{bmatrix}$

$$\text{Directional divergence over sphere surface} = \frac{4\pi a^3 \alpha_y \left(2a^2 + 5\right)}{15}$$

$$\text{Directional divergence over sphere volume} = \frac{4\pi a^3 \alpha_y \left(2a^2 + 5\right)}{15}$$

`The directional divergence formula holds? True`

In this case, the volume integrand becomes quite complex and lengthy (thus, it is not printed). However, the directional divergence formula holds regardless of the values of α_i. Readers may use the provided code to conduct similar examinations with other types of weight and vector functions.

8.5.3 *A directional divergence theorem*

Furthermore, by setting $W = 1$ in Eq. (8.30), and thus $\nabla_\alpha W = 0$, we obtain

$$\oint_S \mathbf{v}_\alpha \cdot d\mathbf{S} = \iiint \text{div}_\alpha \, \mathbf{v} \, dV \tag{8.32}$$

This is a **directional divergence theorem** in 3D, useful in derivative smoothing operations within computational methods, including the S-FEM [9].

For 2D problems, we have

$$\oint_C \mathbf{v}_\alpha \cdot d\mathbf{s} = \iint \text{div}_\alpha \, \mathbf{v} \, dA \tag{8.33}$$

This is the **directional divergence theorem** in 2D.

8.5.4 *Application to the SPH method*

In a local smoothing domain $\mathcal{D}_i$ within a discretized domain if we set

1. the scalar function $W(x, y, z)$ as a bell-shaped weight (or smoothing) function
2. and design the weight function $W(x, y, z)$ so that $W(x, y, z)\big|_S = 0$,

then Eq. (8.5) becomes

$$\underbrace{\frac{1}{V_i} \iiint_{\mathcal{D}_i} W \, \text{div}_\alpha \, \mathbf{v} \, dV}_{\text{smoothed directional divergence}} = -\frac{1}{V_i} \iiint_{\mathcal{D}_i} \mathbf{v} \cdot (\nabla_\alpha W) dV \tag{8.34}$$

This equation can be used to approximate the divergence of the velocity function in the SPH method [5], as discussed earlier. The main difference is that we now have a new parameter, α_i, which provides greater flexibility for constructing more effective or alternative algorithms.

8.6 Concluding remarks

This chapter presents Gauss's formula, a widely used tool in computational methods. Extensions derived from Gauss's formula have also been discussed. We close the chapter with the following remarks:

1. Gauss's formula is an integral relation involving two functions, typically a scalar and a vector function. This provides flexibility for a variety of scalar functions, making it a highly general formula. Theorems discussed in previous chapters can be viewed as special cases of Gauss's formula. Computational methods based on weak and weakened weak forms have largely been developed using Gauss's formula.
2. Gauss's formula extends to an integral relation involving a pair of scalar functions by incorporating directional derivatives. This extension allows the approximation of derivatives or selected derivatives using boundary integrals along curves or surfaces.
3. Gauss's formula also extends, in this chapter, to an integral relation involving a scalar function paired with a vector function through directional divergence. This allows for approximating divergence or partial divergence using boundary integrals along curves or surfaces.

Gauss's formula and its extensions have numerous important applications. These formulas are instrumental in developing computational methods for stress analysis in solids and structures, fluid flow, heat transfer, acoustics, and many other fields. Typical methods include FEM, S-FEM, various mesh-free methods [6, 7], finite volume methods [10], and GSMs [11]. The author believes that these formulas will play a key role in the future development of more advanced and powerful computational methods for a wide range of problems in nature, science, and engineering.

References

[1]   G. Strang, *Calculus*, Wellesley-Cambridge Press, Massachusetts, 1991.
[2]   G.R. Liu, *Numbers and Functions: Theory, Formulation, and Python Codes*, World Scientific, New Jersey, 2024.

[3]   G.R. Liu, *Mechanics of Materials: Formulations and Solutions with Python*, World Scientific, New Jersey, 2024.

[4]   G.R. Liu, *Calculus: A Practical Course with Python*, World Scientific, New Jersey, 2025.

[5]   G.R. Liu and Moubin B. Liu, *Smoothed Particle Hydrodynamics: A Meshfree Particle Method*, World Scientific, 2003.

[6]   G.R. Liu and Y. Gu, *An Introduction to Meshfree Methods and Their Programming*, The Netherlands, 2005.

[7]   G.R. Liu, *Mesh Free Methods: Moving Beyond the Finite Element Method*, Taylor and Francis Group, New York, 2010.

[8]   J. Balaš, C.A. Brebbia, J. Trevelyan *et al.*, *Boundary Element Method: Principles and Applications*, 2000.

[9]   G.R. Liu and T.T. Nguyen, *Smoothed Finite Element Methods*, Taylor and Francis Group, New York, 2010.

[10] R.J. LeVeque, *Finite Volume Methods for Hyperbolic Problems*, Cambridge University Press, Cambridge, 2002.

[11] G.R. Liu and Zirui Mao, *Gradient Smoothing Methods with Programming: Applications to Fluids and Landslides*, World Scientific, New Jersey, 2023.

[12] G.R. Liu and Gui-Yong, Zhang, *Smoothed Point Interpolation Methods: G Space Theory and Weakened Weak Forms*, World Scientific, New Jersey, 2013.

Chapter 9

Conservative and Divergence-Free Fields

```
 1  # Place cursor in this cell, and press Ctrl+Enter to import dependences.
 2  import sys                        # For accessing the computer system
 3  sys.path.append('../grbin/')         # Add in the path to your system
 4
 5  from commonImports import *      # Import dependences from '../grbin/'
 6  import grcodes as gr                # Import the module of the author
 7  importlib.reload(gr)             # When grcodes is modified, reload it
 8
 9  init_printing(use_unicode=True)       # For latex-like quality printing
10  np.set_printoptions(precision=4,suppress=True,
11                      formatter={'float_kind': '{:.4e}'.format})
```

In Chapters 3 and 8, we discussed a special conservative vector field and techniques with code to find its potential function. In Chapter 4, we discussed a special divergence-free vector field and techniques with code to find its stream function. Is there a field that is both conservative and divergence-free (CDF)? The answer is yes; there are many such fields in nature, science, and engineering. In fluid dynamics, for example, there is a class of flows known as ideal flows, which are typically incompressible and irrotational. Such a field has both a potential and a stream function, and is governed by the well-known Laplace equation. This chapter discusses this special field, examines its features, and presents techniques with code to construct CDF fields.

This chapter references textbooks in Refs. [1–3]. Wikipedia pages, particularly those on Cauchy–Riemann equations (https://en.wikipedia.org/wiki/Cauchy%E2%80%93Riemann_equations) and stream function (https://en.wikipedia.org/wiki/Stream_function), serve as valuable additional references. Both NumPy and SymPy are used in the development of code for the demonstration examples. Discussions with ChatGPT, Gemini, and Bing have also greatly helped in coding and in the preparation of this chapter.

9.1 Cauchy–Riemann equations

The condition for a field to be both CDF is that both a potential function and a stream function can be found, and the components of the vector field can be written as

$$
V_x = \boxed{\frac{\partial \phi}{\partial x}} = \frac{\partial \psi}{\partial y}
$$

$$
V_y = \boxed{\frac{\partial \phi}{\partial y}} = -\frac{\partial \psi}{\partial x}
$$

(9.1)

These (boxed) relationships are known as the Cauchy–Riemann equations for the potential function ϕ and stream function ψ.

Since we have already developed Python functions to find potential and stream functions, it is straightforward to examine a given vector field. Let's start the examination with a simple example that is familiar to us.

9.2 Example: A simple CDF field

Consider the simple vector field given by

$$
\mathbf{V}(x, y) = \begin{bmatrix} V_x \\ V_y \end{bmatrix} = \begin{bmatrix} y \\ x \end{bmatrix}
$$

(9.2)

We will show that this field is both CDF, and we will find its potential and stream functions, which obey Eq. (9.1). The code snippets to confirm this are as follows:

```
1  x, y = symbols('x, y', real=True)              # define variables
2  X = sp.Matrix([x, y])
3
4  # An irrotational & incompressible vector field:
5  Vx = y; Vy = x
6
7  ϕ = gr.find_potential_f2D(Vx, Vy, X)
8  display(Math(f" \\text{{The potential function ϕ found}}={latex(ϕ)}"))
9
10 ψ = gr.find_stream_f2D(Vx, Vy, X)
11 display(Math(f" \\text{{The stream function ψ found}}={latex(ψ)}"))
```

```
ɸ = x*y
(ɸ.diff(x)-Vx).simplify() = 0
(ɸ.diff(y)-Vy).simplify() = 0
```

The potential function ϕ found $= xy$

```
ψ = -x**2/2 + y**2/2
(ψ.diff(x)+Vy).simplify() = 0
(ψ.diff(y)-Vx).simplify() = 0
```

The stream function ψ found $= -\dfrac{x^2}{2} + \dfrac{y^2}{2}$

9.3 The Laplace equation

The Laplace equation is one of the most frequently encountered PDEs in calculus. It governs many problems in science and engineering. Here, it can be derived using the properties of a CDF field.

9.3.1 *Vector component relationships*

The CDF vector field given in Eq. (9.1) satisfies the following equations:

$$\boxed{\frac{\partial V_x}{\partial y} = \frac{\partial V_y}{\partial x}}$$

$$\boxed{\frac{\partial V_x}{\partial x} = -\frac{\partial V_y}{\partial y}} \tag{9.3}$$

This defines the **vector component relationships**. It results further in

$$\underbrace{\frac{\partial^2 \phi}{\partial x^2} + \frac{\partial^2 \phi}{\partial y^2}}_{\partial^2 \phi} = \frac{\partial V_x}{\partial x} + \frac{\partial V_y}{\partial y} = 0$$

$$\underbrace{\frac{\partial^2 \psi}{\partial x^2} + \frac{\partial^2 \psi}{\partial y^2}}_{\partial^2 \psi} = \frac{\partial V_x}{\partial y} - \frac{\partial V_y}{\partial x} = 0 \tag{9.4}$$

These are two **Laplace equations**: The first is for the potential function ϕ and the second is for the stream function ψ. These equations can be confirmed for the vector field given in Eq. (9.2) using the following snippet:

```
1  (ɸ.diff(x,2)+ɸ.diff(y,2)),(ψ.diff(x,2)+ψ.diff(y,2))
```

(0, 0)

9.3.2 *Orthogonality of potential and stream functions*

It is straightforward to confirm that the potential function and the stream function for the CDF field given in Eq. (9.2) are orthogonal. This implies that the dot product of the gradient of ϕ and that of ψ vanishes:

```
1  gr.grad_f(ϕ, X).T@gr.grad_f(ψ, X)
```

$[0]$

In fact, the orthogonality of the potential function ϕ and the stream function ψ can be easily proven for any vector field that is both CDF. This is because such a field satisfies the Cauchy–Riemann equations, as given by Eq. (9.1), which can be rewritten as

$$\nabla\phi = \begin{bmatrix} \frac{\partial\phi}{\partial x} \\ \frac{\partial\phi}{\partial y} \end{bmatrix} = \begin{bmatrix} V_x \\ V_y \end{bmatrix}; \quad \nabla\psi = \begin{bmatrix} \frac{\partial\psi}{\partial x} \\ \frac{\partial\psi}{\partial y} \end{bmatrix} = \begin{bmatrix} V_y \\ -V_x \end{bmatrix} \tag{9.5}$$

Therefore, the dot product of $\nabla\phi$ and $\nabla\psi$ vanishes:

$$\nabla\phi \cdot \nabla\psi = V_x V_y - V_y V_x = 0 \tag{9.6}$$

This is the **orthogonality relationship** between the potential function ϕ and the stream function ψ for a CDF field.

9.3.3 *Case study: A potential, incompressible field*

Consider a fluid flow field given by

$$\mathbf{V}(x, y) = \begin{bmatrix} V_x \\ V_y \end{bmatrix} = \begin{bmatrix} -x \\ y \end{bmatrix} \tag{9.7}$$

This is a potential, incompressible field, which is a typical CDF field. The tasks of this study are as follows:

1. Show that the field is irrotational, implying that its curl is zero.
2. Demonstrate that the flow is a potential flow and, therefore, is conservative, implying that a potential function for this field can be found. Then, find the potential function.
3. Show that the field is divergence-free. This means that the fluid is incompressible and that the volume of a fluid cell will not change — it may be compressed in one direction but stretched in another, preserving the volume. Then, find the stream function.
4. Plot the contours of both the potential and stream functions and discuss the results.

Solution: We write the following snippet to perform the tasks:

```
1  # Define the vector field:
2
3  x, y = symbols('x, y', real=True)                    # define variables
4  X = sp.Matrix([x, y])
5
6  # An irrotational and incompressible flow:
7  Vx = -x; Vy = y
8  V = sp.Matrix([Vx, Vy])
9
10 # Answer to question 1:
11 curl_V = gr.curl_vf(V, X).simplify()             # the curl of the field
12 display(Math(f" \\text{{ curl of the field}}={latex(curl_V)}"))
13
14 # Answer to question 2:
15 ϕ = gr.find_potential_f2D(Vx, Vy, X)
16 display(Math(f" \\text{{The potential function ϕ found = }}{latex(ϕ)}"))
```

Curl of the field $= 0$

```
ϕ = -x**2/2 + y**2/2
(ϕ.diff(x)-Vx).simplify() = 0
(ϕ.diff(y)-Vy).simplify() = 0
```

The potential function ϕ found $= -\dfrac{x^2}{2} + \dfrac{y^2}{2}$

The curl of this field is found to be zero, indicating that the flow field is irrotational, which answers question 1.

Additionally, the field is a potential flow and is conservative, and the potential function has been found, addressing question 2.

Solution to Question 3:

```
1  # Answer to question 3-1:
2  div_V = gr.div_vf(V, X)
3  display(Math(f" \\text{{ Divergence of the field}}={latex(div_V)}"))
4
5  # Answer to question 3-2:
6  ψ = gr.find_stream_f2D(Vx, Vy, X)
7  display(Math(f" \\text{{ The stream function ψ found}}={latex(ψ)}"))
```

Divergence of the field $= 0$

```
ψ = -x*y
(ψ.diff(x)+Vy).simplify() = 0
(ψ.diff(y)-Vx).simplify() = 0
```

The stream function ψ found $= -xy$

This completes the solution to Question 3.

Solution to Question 4: Finally, let's plot the contour lines of both the potential function and the stream function for this flow field. For convenience in future use, we've written the following Python function, which can plot the contours of both the potential and stream functions for a given CDF vector field that is both CDF. The function plots the upper part of the field:

```python
def draw_Ellipse(center, a, b, θ, lw=2, alpha=0.6):
    '''Draw an ellipse and add it to a plot, with the following
       parameters (with examples):
       center = (0, 0)    # Center of the ellipse
       a  = 1.5           # Major axis length
       b  = 0.5           # Minor axis length
       θ  = 30            # Rotation angle in degrees
       lw = 2             # line width
       alpha              # transparent factor
       needs a "from matplotlib.patches import Ellipse"
       Written based on a suggestion by ChatGPT.
    '''
    # Add an ellipse to the plot
    ellipse = Ellipse(xy=center, width=a, height=b, angle=θ, \
                      edgecolor='g', fc='g', lw=lw, alpha=alpha)
    plt.gca().add_patch(ellipse)

    # Calculate the end points of the major and minor axes
    theta = np.radians(θ)
    major_axis_length = a / 2
    minor_axis_length = b / 2

    # For drawing the major axis
    major_x = [center[0] - major_axis_length * np.cos(theta), \
               center[0] + major_axis_length * np.cos(theta)]
    major_y = [center[1] - major_axis_length * np.sin(theta), \
               center[1] + major_axis_length * np.sin(theta)]

    # For drawing the minor axis
    minor_x = [center[0] - minor_axis_length * np.sin(theta), \
               center[0] + minor_axis_length * np.sin(theta)]
    minor_y = [center[1] + minor_axis_length * np.cos(theta), \
               center[1] - minor_axis_length * np.cos(theta)]

    plt.plot(major_x, major_y, 'b', label='Major Axis')
    plt.plot(minor_x, minor_y, 'b--', label='Minor Axis')
```

```
 1  def plot_potentialStream_f(ϕ, ψ, C, xL, xR, yL, yU, contourδ=0.0005):
 2      '''Plots the contours of both potential ϕ and stream ψ functions
 3          all in sympy, in the domain of [xL, xR]×[yL, yU].
 4          contourδ is used put the arrows of the tangential direction of
 5          the streamfunction ψ.
 6      '''
 7      plt.figure(figsize=(7, 4))                              # in inches
 8      c = sp.symbols("c", real=True)
 9      ϕ = ϕ - c                               # potential functions
10      ψ = ψ - c                               # stream functions
11
12      n_points = 200                          # Create mesh grid
13      x_ = np.linspace(xL, xR, n_points)
14      y_ = np.linspace(yL, yU, n_points)
15      X_, Y_ = np.meshgrid(x_, y_)
16
17      for c_ in C:                            # Loop over all contour constants
18          np_ϕ = lambdify((x,y), ϕ.subs(c,c_), 'numpy') # convert f to np
19          np_ψ = lambdify((x,y), ψ.subs(c,c_), 'numpy')
20          Zϕ = np_ϕ(X_, Y_)                   # Compute Z values over the mesh
21          Zψ = np_ψ(X_, Y_)
22
23          # Compute the gradient of ψ: ∇ψ=[dψ/dx, dfψdy], & covert to np
24          dψdx = lambdify((x,y), (ψ.diff(x)).subs(c,c_), 'numpy')
25          dψdy = lambdify((x,y), (ψ.diff(y)).subs(c,c_), 'numpy')
26
27          # Compute the tangential vector t (-dfdy, dfdx), ⊥ to ∇ψ
28          Tψx = -dψdy(X_, Y_); Tψy = dψdx(X_, Y_)
29
30          magnitude = np.sqrt(Tψx**2 + Tψy**2)         # unit vector t
31          Tψx /= magnitude;    Tψy /= magnitude
32
33          mask = np.abs(Zψ) < contourδ
34          plt.contour(X_, Y_, Zϕ, levels=[0], linewidths=0.5, colors='m')
35          plt.contour(X_, Y_, Zψ, levels=[0], linewidths=0.5, colors='b')
36
37          # Plot arrows in tangential direction:
38          plt.quiver(X_[mask], Y_[mask], Tψx[mask], Tψy[mask], \
39                              color='b', scale=60, width=0.01)
40
41      plt.axis('equal') #plt.xlim(-2, 2); plt.ylim(0.5, 1.5)
42      plt.axvline(x=0, c="k", lw=0.6); plt.axhline(y=0, c="k", lw=0.2)
43      plt.xlabel('x'); plt.ylabel('y'); #plt.grid(True)
```

The following snippet plots the contour of the potential function ϕ along with that of the stream function ψ, with arrows indicating the tangential direction along the contour lines:

```
 1  from matplotlib.patches import Ellipse  # for plotting ellipses on plot
 2
 3  x, y = sp.symbols("x, y",real=True)
 4  φ = -x**2/2 + y**2/2
 5  ψ = -x*y
 6  display(Math(f" \\text{{Potential function = }} {latex(φ)}; \\;\\;\\;\
 7                     \\text{{Stream function = }} {latex(ψ)}"))
 8
 9  Vx = φ.diff(x); Vy = φ.diff(y);
10  display(Math(f" \\text{{Flow field via potential function $V_{{x}}$=}}\
11             {latex(Vx)}; \\;\\;\\; \\text{{ $V_{{y}}$ = }}{latex(Vy)}"))
12
13  Vx = ψ.diff(y); Vy = -ψ.diff(x);
14  display(Math(f" \\text{{Flow field via stream function $V_{{x}}$ = }}\
15             {latex(Vx)}; \\;\\;\\; \\text{{ $V_{{y}}$ = }}{latex(Vy)}"))
16
17  plt.ioff() # plt.ion()          # use plt.ion() to generate a new plot
18  C = np.arange(-2, 2.0, 0.2)              # constants for contour levels
19  xL = -2; xR = 2
20  yL = 0.01; yU = 2
21  plot_potentialStream_f(φ, ψ, C, xL, xR, yL, yU, contourδ=0.0005)
22
23  # Add an elliptic cell in the irrotational and incompressible flow:
24  a = 0.5; b = 0.2
25  r = np.sqrt(a*b)                      # to ensure the area is the same
26  draw_Ellipse((0.4, 1.5), r, r, 0, lw=2, alpha=0.6) # a circle of A=πr^2
27  draw_Ellipse((1.2, 0.5), a, b, 0, lw=2, alpha=0.6) # an ellipse, same A
28
29  plt.scatter((0),(0), color='r', s=30, zorder=3)       # A saddle points
30  plt.savefig('imagesVC/potentialStream_z2.png', dpi=500)  # save to file
31  #plt.show()
```

Potential function $= -\dfrac{x^2}{2} + \dfrac{y^2}{2};$ Stream function $= -xy$

Flow field via potential function $V_x = -x;$ $V_y = y$

Flow field via stream function $V_x = -x;$ $V_y = y$

The flow field computed using the potential and stream functions is the same. The plot for both functions is given as follows.

We find the following:

1. It can be observed from Fig. 9.1 that the contour lines of the potential function $\phi(x, y) =$ constants, and those of the stream function $\psi(x, y) =$ constants, are orthogonal to each other at any point in the field.

2. The potentials drive the fluid flows along the streamlines.

3. Since the field is divergence-free, the volume of a cell in the fluid will not change. For example, a circle upstream may change to an ellipse downstream, but its area will remain constant.

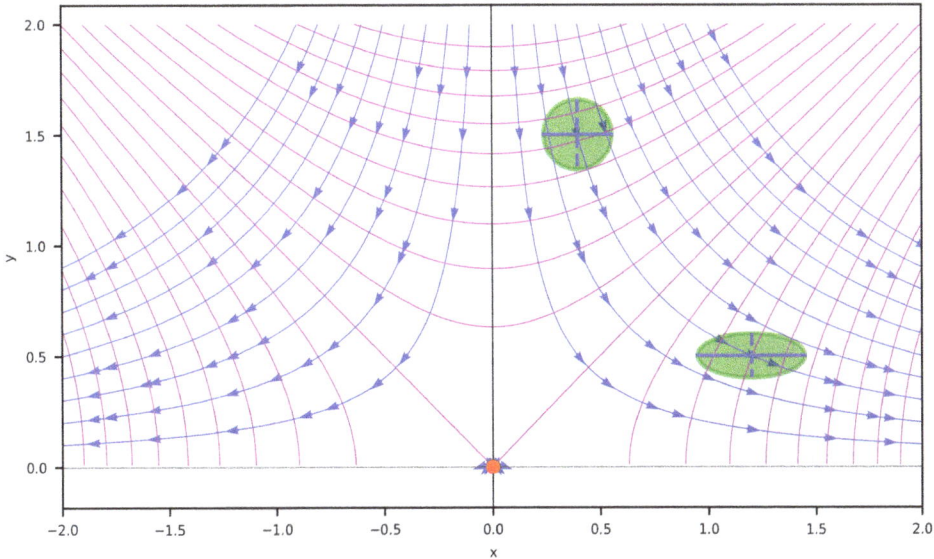

Figure 9.1. Schematic view of the potential function (magenta) and the stream function (blue) for the vector field $[-x \quad y]^\top$. These contour lines are orthogonal to each other at any point in the domain.

4. Since the field is irrotational, the orientation of the circle upstream remains the same as that of the ellipse downstream. No rotation is observed.

5. At the origin $(0,0)$, we observe that the fluid flows into it in the vertical direction but flows out from it in the horizontal direction. This is a typical saddle point, as discussed in Ref. [4].

The features of a vector field at a point can be analyzed using its partial derivatives. The first approach uses the Jacobian matrix, which consists of the partial derivatives of the component functions.

Analysis using the Jacobian matrix: The Jacobian matrix of a vector field at a point provides insight into how the vector field changes near that point, as discussed in Section 2.12.3.

Let us compute the Jacobian matrix for the potential, incompressible flow field given in Eq. (9.7):

```
1  # Generate the Jacobian matrix, using derivatives of the vector field:
2
3  J = sp.simplify(gr.grad_vf(Matrix([Vx, Vy]), X).T)   # need a transpose
4  display(Math(f" \\text{{Jacobian matrix J = }}{{latex(J)}}"))
```

Jacobian matrix $\mathbf{J} = \begin{bmatrix} -1 & 0 \\ 0 & 1 \end{bmatrix}$

This matrix is simple, and its eigenvalues appear directly on the diagonal. One eigenvalue is positive, and the other is negative. Therefore, every point in the field is a saddle point. As a typical example, the red dot marked in Fig. 9.1 clearly indicates a saddle point: The fluid flows into the origin vertically and flows out horizontally.

Analysis using the Hessian matrix: Alternatively, we can analyze the Hessian matrix of the scalar potential function similar to the method described in Ref. [4]. For this particular case, the Hessian matrix for the scalar potential function ϕ is found as

$$\mathbf{H}(\phi) = \begin{bmatrix} \frac{\partial^2 \phi}{\partial x^2} & \frac{\partial^2 \phi}{\partial x \partial y} \\ \frac{\partial^2 \phi}{\partial y \partial x} & \frac{\partial^2 \phi}{\partial y^2} \end{bmatrix} \tag{9.8}$$

We can then compute the determinant of $\mathbf{H}$, $\det(\mathbf{H})$, and the trace of $\mathbf{H}$, $\text{tr}(\mathbf{H})$, at a point.

- If $\det(\mathbf{H}) > 0$ and $\text{tr}(\mathbf{H}) > 0$, the point is a local minimum of the potential function. The corresponding vector field at the point is repulsive.
- If $\det(\mathbf{H}) > 0$ and $\text{tr}(\mathbf{H}) < 0$, the point is a local maximum of the potential function. The corresponding vector field at the point is attractive.
- If $\det(\mathbf{H}) < 0$, the point is a saddle point. The corresponding vector field flows into the point in some directions and out from it in others.

Let us compute the Hessian matrix for the potential, incompressible flow field given in Eq. (9.7):

```
1  Hessian = sp.hessian(ϕ, X)
2  Hessian
```

$\begin{bmatrix} -1 & 0 \\ 0 & 1 \end{bmatrix}$

Both the determinant and trace of the Hessian matrix are trivial, indicating that any point in this field is a saddle point.

In fact, for any potential incompressible flow field, the Jacobian and Hessian matrices are essentially the same. It is a good exercise for the reader to show that this is true.

9.3.4 *Example: Gravity field — CDF*

This is another typical CDF field. We studied earlier that the gravity field is conservative. We will show here that it is also divergence-free. We can find both its potential and stream functions.

Let us use `find_potential_f2D()` to find the potential function for the gravity field:

```
1  m, g, H = symbols('m, g, H', positive=True)
2  Gx = 0*y; Gy = -m*g                          # the gravity field
3  φ = gr.find_potential_f2D(Gx, Gy, X)
4  gr.printM(φ, ' The potential function φ found is: ')
```

```
φ = -g*m*y
(φ.diff(x)-Vx).simplify() = 0
(φ.diff(y)-Vy).simplify() = 0
 The potential function φ found is
```

$$-gmy$$

Next, use `find_stream_f2D()` to find the stream function for the gravity field:

```
1  ψ = gr.find_stream_f2D(Gx, Gy, X)
2  gr.printM(ψ, ' The stream function ψ found is: ')
```

```
ψ = g*m*x
(ψ.diff(x)+Vy).simplify() = 0
(ψ.diff(y)-Vx).simplify() = 0
 The stream function ψ found is:
```

$$gmx$$

As seen in the gravity field example, the contour lines of the potential function are horizontal lines, while those of the stream function are vertical lines similar to the lines representing falling rain. This provides a clear intuition about the physical meaning of these two functions. The horizontal lines representing the potentials of the gravity field pull the water droplets vertically downward.

Let us take a look at the Jacobian matrix for the gravity field:

```
1  J = sp.simplify(gr.grad_vf(Matrix([Gx, Gy]), X).T)    # need a transpose
2  display(Math(f" \\text{{Jacobian matrix J = }}{latex(J)}"))
```

Jacobian matrix $\mathbf{J} = \begin{bmatrix} 0 & 0 \\ 0 & 0 \end{bmatrix}$

It is a zero matrix because the gravity field is a constant field. The derivatives of its components are all zero.

9.4 Complex functions for CDF fields

9.4.1 *Analytic functions and their properties*

A complex function is said to be **analytic** at a point if it can be locally expressed by a convergent power series. Thus, an analytic function is complex differentiable at that point and in a neighborhood around it. An analytic function has several important properties, two of which are relevant to our discussion here. The first is that the real and imaginary parts of the function satisfy the Cauchy–Riemann equations.

Consider a general analytic function defined as

$$f(z) = u(x, y) + iv(x, y) \tag{9.9}$$

where $z = x + iy$ with i being the imaginary unit, $i = \sqrt{-1}$. We then have

$$\frac{\partial u}{\partial x} = \frac{\partial v}{\partial y}, \quad \frac{\partial u}{\partial y} = -\frac{\partial v}{\partial x} \tag{9.10}$$

which is exactly the same as the relations given in Eq. (9.1) if we let $\phi = u$ and $\psi = v$!

Equation (9.10) provides the necessary conditions, and when these partial derivatives are all continuous, the complex function $f(z)$ will be analytic.

Secondly, an analytic function is also holomorphic, implying that the real and imaginary parts of an analytic function are harmonic functions. This means that they satisfy the Laplace equation: $\nabla^2 u = 0$ and $\nabla^2 v = 0$. This is also shown in Eq. (9.4).

9.4.2 *List of typical analytic functions*

Examples of complex analytic functions are given as follows:

1. **Polynomial functions** (analytic over the entire complex plane):

$$f(z) = c_0 + c_1 z + c_2 z^2 + \cdots + c_n z^n \qquad (9.11)$$

where c_i for $i = 0, \ldots, n$ are complex constants.

2. **Power functions** (with noninteger power c): z^c
   This function is analytic in regions that avoid branch points (typically along a branch cut on the negative real axis).

3. **Rational functions:** $\frac{P(z)}{Q(z)}$
   where $P(z)$ and $Q(z)$ are polynomials, and $Q(z) \neq 0$ in the domain of interest. Rational functions are analytic except at poles where $Q(z) = 0$.

4. **Exponential function** (analytic over the entire complex plane): e^z

5. **Logarithmic function** (inverse of the exponential function): $\log(z)$
   It is analytic in regions that avoid the branch cut.

6. **Trigonometric functions** (analytic over the entire complex plane): $\sin(z), \cos(z)$

7. **Inverse trigonometric functions:** $\arcsin(z), \arccos(z), \arctan(z)$
   These functions are analytic in regions that avoid their branch points.

8. **Hyperbolic functions** (analytic over the entire complex plane): $\sinh(z), \cosh(z)$
   These functions are made of the exponential function.

9.4.3 *Construction of potential and stream functions*

With this understanding, we can utilize the properties of analytic functions to generate potential and stream functions. To do this, we need to create an analytic function and then split it into its real and imaginary parts. Splitting a complex function into real and imaginary parts in closed form is not always straightforward but can be done in many cases [2]. When such a splitting is possible, the analytic function becomes a complex potential function that provides closed forms for both potential and stream functions for a vector field.

Following is an example to demonstrate this process using a simple complex function:

$$f(z) = z^2 \tag{9.12}$$

This function is a single term from the complex polynomial function and is analytic. We can use its real part as the potential function and the imaginary part as the stream function (or vice versa). This process is straightforward, and the snippet is as follows:

```
1  x, y = sp.symbols("x, y",real=True)
2  X = sp.Matrix([x, y])
3  z = x + sp.I*y                    # sp.I: imaginary number defined in SymPy.
4
5  fz = -z**2/2                              # Complex potential function
6  ϕ = sp.re(fz)                            # get the real part of fz
7  ψ = sp.im(fz)
8  orth_ϕψ = gr.grad_f(ϕ,X).T@gr.grad_f(ψ,X)
9  display(Math(f" \\text{{Orthogonal?}} \\; {latex(orth_ϕψ[0]==0)}"))
10
11 display(Math(f" \\text{{Potential function: Real part of}} {latex(fz)}\
12                                      = {latex(ϕ)}"))
13 display(Math(f" \\text{{Stream function: Imaginary part of}}{latex(fz)}\
14                                      = {latex(ψ)}"))
15 Vx = ψ.diff(y); Vy = -ψ.diff(x);
16 display(Math(f" \\text{{Flow field $V_{{x}}$ = }}{latex(Vx)}; \\;\\;\\;\
17                 \\text{{ $V_{{y}}$ = }}{latex(Vy)}"))
```

Orthogonal? True

Potential function: Real part of $-\dfrac{(x+iy)^2}{2} = -\dfrac{x^2}{2} + \dfrac{y^2}{2}$

Stream function: Imaginary part of $-\dfrac{(x+iy)^2}{2} = -xy$

Flow field $V_x = -x; \quad V_y = y$

We successfully split this complex potential function and obtained both the potential and stream functions of this simple CDF flow. The results are exactly the same as those obtained in Section 9.3.3.

9.4.4　A periodic vector field of CDF flow

Let us now create a new field using a complex sine function:

Potential function: Real part of $\sin(\omega(x+iy)) = \sin(x\omega)\cosh(y\omega)$

```
1   x, y = sp.symbols("x, y",real=True)
2   z = x + sp.I*y                      # sp.I: imaginary number defined in SymPy.
3
4   ω = sp.symbols("ω",real=True)        # a parameter for the sine function
5   fz = sin(ω*z)                        # Complex potential function, it is analytic
6
7   φ = sp.re(fz)                                    # get the real part of fz
8   ψ = sp.im(fz)
9   display(Math(f" \\text{{Potential function: Real part of}} {latex(fz)}\
10                                           = {latex(φ)}"))
11  orth_φψ = gr.grad_f(φ,X).T@gr.grad_f(ψ,X)
12  display(Math(f" \\text{{Orthogonal?}} \\; {latex(orth_φψ[0]==0)}"))
13
14  display(Math(f" \\text{{Stream function: Imaginary part of}}{latex(fz)}\
15                                           = {latex(ψ)}"))
16  Vx = φ.diff(x); Vy = φ.diff(y)
17  display(Math(f" \\text{{Flow field via φ, $V_{{x}}$ = }}{latex(Vx)};\
18                      \\;\\;\\; \\text{{ $V_{{y}}$ = }}{latex(Vy)}"))
19  Vx = ψ.diff(y); Vy = -ψ.diff(x)
20  display(Math(f" \\text{{Flow field via ψ, $V_{{x}}$ = }}{latex(Vx)};\
21                      \\;\\;\\; \\text{{ $V_{{y}}$ = }}{latex(Vy)}"))
```

Orthogonal? True

Stream function: Imaginary part of $\sin(\omega(x+iy)) = \cos(x\omega)\sinh(y\omega)$

Flow field via ϕ, $V_x = \omega\cos(x\omega)\cosh(y\omega)$; $V_y = \omega\sin(x\omega)\sinh(y\omega)$

Flow field via ψ, $V_x = \omega\cos(x\omega)\cosh(y\omega)$; $V_y = \omega\sin(x\omega)\sinh(y\omega)$

Again, we successfully split this complex potential function and obtained both the potential and stream functions of a CDF flow. Computing the component functions for this new vector field is straightforward, as it involves only partial differentiation and can always be done.

Let us check whether they satisfy the Laplace equations:

```
1   (φ.diff(x,2)+φ.diff(y,2)), (ψ.diff(x,2)+ψ.diff(y,2))
```

(0, 0)

Yes, they do. This field is sinusoidal. The periodicity is $\frac{2\pi}{\omega}$.

Let us plot the vector field using the same Python function written earlier:

```
1   ω_ = 1.0                                # set the parameter to a fixed value
2   φ = φ.subs(ω,ω_)                               # potential function
3   ψ = ψ.subs(ω,ω_)                                # stream function
4
5   plt.ioff()
6   #plt.ion()                             # use plt.ion() to generate a new plot
7   C = np.arange(-2, 2.0, 0.2)              # contours for potential & stream f
8   xL = -3.2; xR = 3.2
9   yL = 0.01; yU = 4.5
10  plot_potentialStream_f(φ, ψ, C, xL, xR, yL, yU, contourδ=0.0005)
11
12  # Two saddle points:
13  plt.scatter((-np.pi/2, np.pi/2,),(0., 0.), color='r', s=30, zorder=3)
14
15  # Two black holes:
16  plt.scatter((-np.pi/2, np.pi/2,),(yU, yU), color='k', s=60, zorder=3)
17  plt.savefig('imagesVC/potentialStream_sinz.png', dpi=500) #save to file
18
19  #plt.show()
```

We have found a new potential incompressible vector field. Both the potential and stream functions for this vector field are plotted in Fig. 9.2. It has the following features:

1. The contour lines of the potential and stream functions are orthogonal to each other at any point in the domain.

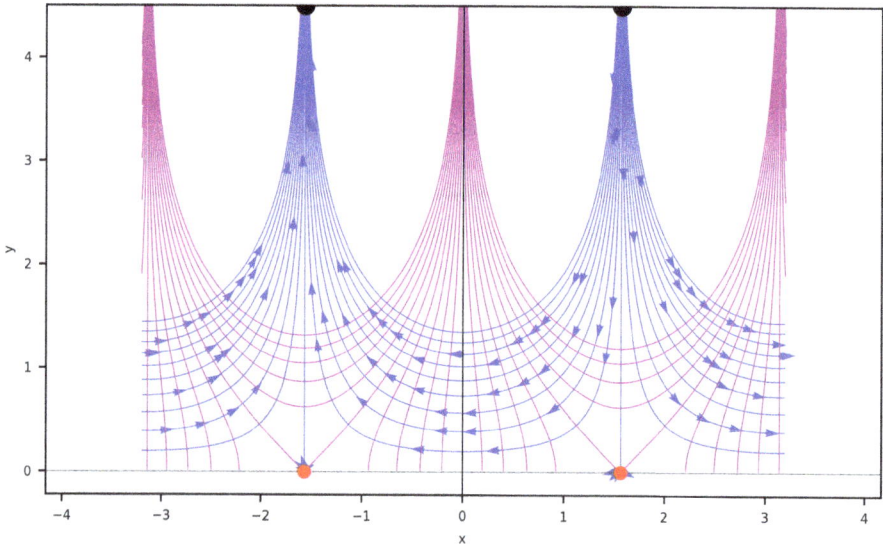

Figure 9.2. Schematic view of the potential function (magenta) and stream function (blue) for the vector field generated by a complex potential that is a simple sine function of z.

2. There are two saddle points marked in red. The fluid flows into the point in one direction and out from it in another direction.
3. These functions are periodic. In the plotted domain, there are two black holes marked with black dots: one on the left swallows fluid and one on the right spits out fluid.

Let us check the eigenvalues of the Jacobian matrix for this potential flow field:

```
1  J = sp.simplify(gr.grad_vf(Matrix([Vx, Vy]), X).T)     # Jacobian matrix
2  display(Math(f" \\text{{Jacobian matrix J = }}{latex(J)}"))
3  eigsJ = J.eigenvals()                      # compute the eigenvalues
4  eigsJ = [e for e in eigsJ]                  # convert the solution to list
5  # Two eigenvalues:
6  display(Math(f" \\text{{1st eigenvalue of J = }}{latex(eigsJ[0])}"))
7  display(Math(f" \\text{{2nd eigenvalue of J = }}{latex(eigsJ[1])}"))
```

$$\text{Jacobian matrix } \mathbf{J} = \begin{bmatrix} -\omega^2 \sin(x\omega)\cosh(y\omega) & \omega^2 \cos(x\omega)\sinh(y\omega) \\ \omega^2 \cos(x\omega)\sinh(y\omega) & \omega^2 \sin(x\omega)\cosh(y\omega) \end{bmatrix}$$

$$\text{First eigenvalue of } \mathbf{J} = -\omega^2 \sqrt{\sin^2(x\omega)\cosh^2(y\omega) + \cos^2(x\omega)\sinh^2(y\omega)}$$

$$\text{Second eigenvalue of } \mathbf{J} = \omega^2 \sqrt{\sin^2(x\omega)\cosh^2(y\omega) + \cos^2(x\omega)\sinh^2(y\omega)}$$

It is observed that we obtain two real eigenvalues: one positive and one negative. This is true for any point in the entire field. This indicates that every point in this vector field is a saddle point. The two red dots marked in Fig. 9.2 are clear examples. As shown in the figure, at any point on a streamline, the fluid flows along it. At the same time, there must be fluid flow toward that point on the streamline in the perpendicular direction due to the irrotational and incompressible nature of the field.

Readers may verify the determinant of the vector function and the Hessian matrix of the potential function using the codes provided earlier.

Readers are also encouraged to generate new CDF flow fields using other analytic functions with the provided code. Care should be taken to avoid singular points (poles) if they are not removable.

9.5 Jacobian matrix for CDF fields

From the previous examples, we found that the two eigenvalues of the Jacobian matrix for a CDF field are equal but with opposite signs. This is not a coincidence and can be theoretically proven.

Since a CDF field satisfies Eq. (9.3), its Jacobian matrix becomes

$$J_{CDF} = \begin{bmatrix} \frac{\partial V_x}{\partial x} & \frac{\partial V_x}{\partial y} \\ \frac{\partial V_x}{\partial y} & -\frac{\partial V_x}{\partial x} \end{bmatrix} \tag{9.13}$$

The eigenvalues are as follows:

$$\lambda_1 = -\sqrt{\left(\frac{\partial V_x}{\partial x}\right)^2 + \left(\frac{\partial V_x}{\partial y}\right)^2}$$

$$\lambda_2 = \sqrt{\left(\frac{\partial V_x}{\partial x}\right)^2 + \left(\frac{\partial V_x}{\partial y}\right)^2} \tag{9.14}$$

which can be obtained using the following code snippet:

```
1  Vx = sp.Function("V_x")(x, y)
2  J_CDF = sp.Matrix([[Vx.diff(x), Vx.diff(y)],
3                     [Vx.diff(y),-Vx.diff(x)]])
4  #display(Math(f" \\text{{Jacobian matrix Jvx = }}{latex(J_CDF)}"))
5  J_CDF.eigenvals()
```

$$\left\{ -\sqrt{\left(\frac{\partial}{\partial x}V_x(x,y)\right)^2 + \left(\frac{\partial}{\partial y}V_x(x,y)\right)^2} : 1, \right.$$

$$\left. \sqrt{\left(\frac{\partial}{\partial x}V_x(x,y)\right)^2 + \left(\frac{\partial}{\partial y}V_x(x,y)\right)^2} : 1 \right\}$$

It is clear from Eq. (9.14) that these two eigenvalues are equal in value but with opposite signs regardless of the value of V_x that may vary over the domain of the vector function. When V_x is a constant, these two eigenvalues become zero, which is the case for the gravity field studied earlier. We note the following as this is an important property:

> Any point in a CDF field is a saddle point unless the field is a constant field.

9.6 A general method for constructing CDF fields

This section proposes a general method for constructing CDF fields. The key ideas for this method are as follows:

1. **Generate harmonic basis functions:** Create the initial form of the potential and stream functions using harmonic basis functions, as they

both satisfy the Laplace equation. The initial form of the potential and stream functions can be expressed as linear combinations of these basis functions along with a set of unknown constants. This follows the general procedure for function construction outlined in Ref. [2].

2. **Determine unknown constants:** Use the vector field component relations for CDF fields, as given in Eq. (9.3), to determine these unknown constants. This results in general forms of the potential and stream functions.

3. **Set constants for particular CDF fields:** By assigning specific values to the constants in the general forms of the potential and stream functions, one can produce a particular CDF field.

9.6.1 *Harmonic basis functions*

Any harmonic basis function satisfies the Laplace equation. Some of these basis functions are provided in Ref. [5]. Following are some harmonic basis functions defined in 2D:

1. **Polynomial harmonic bases:**

$$1, \quad x, \quad y, \quad xy, \quad x^2 - y^2 \leq \quad \text{second order}$$

$$x^2 y - \frac{y^3}{3}, \quad x^3 - 3xy^2 \quad \text{third order}$$

$$x^3 y - xy^3, \quad x^4 - 6x^2 y^2 + y^4 \quad \text{fourth order}$$

$$x^4 y - 2x^2 y^3 + \frac{y^5}{5}, \quad x^5 - 10x^3 y^2 + 5xy^4 \quad \text{fifth order}$$

$$x^5 y - \frac{10x^3 y^3}{3} + xy^5, \quad x^6 - 15x^4 y^2 + 15x^2 y^4 - y^6 \quad \text{sixth order}$$

$$\cdots \tag{9.15}$$

- These polynomial bases are generated using **gr.create_nbf()**, as developed in Ref. [5]. One can generate additional bases using **gr.create_nbf()** if needed.
- These polynomial bases can also be obtained using complex potential functions: $\text{Im}(z^n)$ and $\text{Re}(z^n)$, where $z = x + iy$ and n is an integer. All these will be illustrated in the example section.

Another useful tool is **gr.nabla_f()**, which can be used to check whether a function satisfies the Laplace equation, including the polynomial bases and any other bases mentioned in the following:

2. **Rational harmonic basis:**

$$\frac{x}{x^2 + y^2}, \quad \frac{y}{x^2 + y^2} \tag{9.16}$$

These two form a pair for potential and stream functions.

3. **Logarithmic harmonic basis:**

$$\ln(x^2 + y^2), \quad i\log(x - iy) - i\log(x + iy) \tag{9.17}$$

These two form a pair for potential and stream functions. Note that the second one is complex.

4. **Hyperbolic harmonic basis:**

$$\begin{aligned} \cos(xw)\cosh(yw), \quad &\cos(yw)\cosh(xw), \\ \sin(xw)\cosh(yw), \quad &\cos(yw)\sinh(xw) \end{aligned} \tag{9.18}$$

The sine and cosine functions, as well as the sinh and cosh functions, can also be paired.

5. **Power-law type basis:**

$$|z|^n \cos(n\theta), \quad |z|^n \sin(n\theta) \tag{9.19}$$

where $z = x + iy$ and $\theta = \arg(z)$ for n a positive real number not equal to 1.

These functions are widely used in applications such as fluid flow, electrostatics, and gravitational potentials, where solutions to Laplace's equation describe steady-state potential fields.

One may use a linear combination of some of these basis functions to form a potential or a stream function. Clearly, if the hyperbolic harmonic bases are used, we should be able to produce the CDF field generated using the complex function given earlier.

9.6.2 *Example: Generating polynomial harmonic basis functions*

The following snippet generates the seventh-order polynomial harmonic basis function using our `gr.create_nbf()` method:

```
1  x, y = sp.symbols("x, y", real = True)
2  X = sp.Matrix([x, y])
3
4  order = 7                    # Create 7th-order polynomial harmonic bases
5  H_bases = gr.create_nbf(order, X, dfn=2, prnt=False)
6  display(Math(f"\\text{{Polynomial harmonic bases=}}{latex(H_bases.T)}"))
7  nabla2_Hbases = gr.nabla_f(H_bases, X, dfn=2, smplfy=1)
8  display(Math(f"\\text{{$∇^2=0$?}}{latex(nabla2_Hbases.T)}"))
```

Polynomial harmonic bases

$$= \left[x^6 y - 5x^4 y^3 + 3x^2 y^5 - \tfrac{y^7}{7} \quad x^7 - 21x^5 y^2 + 35x^3 y^4 - 7xy^6 \right]$$

$$\nabla^2 = 0? \begin{bmatrix} 0 & 0 \end{bmatrix}$$

Let us use the complex harmonic basis functions of the same order to produce the basis function:

```
1  n = sp.symbols("n", integer = True)
2
3  # define the potential function
4  φ = sp.re((x+sp.I*y)**7)
5  display(Math(f" \\text{{The potential function φ}}={latex(φ)}"))
6
7  nabla2_φ = gr.nabla_f(φ, X, dfn=2, smplfy=1).simplify()   # Check ∇^2=0?
8  display(Math(f" \\text{{Does φ satisfy $∇^2φ$=0?}}=\
9                            {latex(nabla2_φ==0)}"))
10 Vx = φ.diff(x).simplify(); Vy = φ.diff(y).simplify()   # vector field
11 display(Math(f" \\text{{Vector field found via φ}}={latex((Vx, Vy))}"))
```

The potential function $\phi = x^7 - 21x^5 y^2 + 35x^3 y^4 - 7xy^6$

Does ϕ satisfy $\nabla^2\phi$=0? True

Vector field found via ϕ

$$= \left(7x^6 - 105x^4 y^2 + 105x^2 y^4 - 7y^6, \ -42x^5 y + 140x^3 y^3 - 42xy^5 \right)$$

The complex potential function produces one of the real polynomial basis functions generated previously (up to an immaterial coefficient). However, to obtain the other basis function, we need to use the imaginary part of the complex potential function:

```
1  # Use the complex potential function for the stream function:
2  ψ = sp.im((x+sp.I*y)**7)                           # notice the minus sign!
3  display(Math(f" \\text{{The potential function ψ}}={latex(ψ)}"))
4
5  nabla2_ψ = gr.nabla_f(ψ, X, dfn=2, smplfy=1).simplify()   # Check ∇^2=0?
6  display(Math(f" \\text{{Does ψ satisfy $∇^2ψ$=0?}}=\
7                                              {latex(nabla2_ψ==0)}"))
8  Vxψ = ψ.diff(y).simplify(); Vyψ =-ψ.diff(x).simplify()   # vector field
9  display(Math(f"\\text{{Vector field found via ψ}}={latex((Vxψ,Vyψ))}")
```

The potential function $\psi = 7x^6 y - 35x^4 y^3 + 21x^2 y^5 - y^7$

Does ψ satisfy $\nabla^2\psi=0$? True

Vector field found via ψ

$$= \left(7x^6 - 105x^4 y^2 + 105x^2 y^4 - 7y^6, \ -42x^5 y + 140x^3 y^3 - 42xy^5\right)$$

Alternatively, we can use the vector function obtained via a basis function to generate the other function (the stream function) using our `gr.find_stream_f2D()` method:

```
1  ψ = gr.find_stream_f2D(Vx, Vy, X)
2  display(Math(f" \\text{{The stream function ψ found}}={latex(ψ)}"))
3  nabla2_ψ = gr.nabla_f(ψ, X, dfn=2, smplfy=1)              # Check ∇^2=0?
4  display(Math(f" \\text{{Does ψ satisfy $∇^2ψ$=0?}}=\
5                                              {latex(nabla2_ψ==0)}"))
6  Vx = ψ.diff(y).simplify(); Vy =-ψ.diff(x).simplify()      # vector field
7  display(Math(f" \\text{{Vector field found via ψ}}={latex((Vx, Vy))}"))
```

ψ = y*(7*x**6 - 35*x**4*y**2 + 21*x**2*y**4 - y**6)
(ψ.diff(x)+Vy).simplify() = 0
(ψ.diff(y)-Vx).simplify() = 0

The stream function ψ found $= y\left(7x^6 - 35x^4 y^2 + 21x^2 y^4 - y^6\right)$

Does ψ satisfy $\nabla^2\psi=0$? True

Vector field found via

$$\psi = \left(7x^6 - 105x^4 y^2 + 105x^2 y^4 - 7y^6, \ -42x^5 y + 140x^3 y^3 - 42xy^5\right)$$

We obtained the same results. Let us check the eigenvalues of its Jacobian matrix:

```
1  J_7th = sp.Matrix([[Vx.diff(x), Vx.diff(y)],
2                      [Vx.diff(y),-Vx.diff(x)]])
3  display(Math(f" \\text{{Jacobian matrix of 7th-order bases =}}\
4                                      \\small {latex(J_7th)}"))
5  J_7th.eigenvals()
```

Jacobian matrix of seventh-order bases

$$= \begin{bmatrix} 42x^5 - 420x^3y^2 + 210xy^4 & -210x^4y + 420x^2y^3 - 42y^5 \\ -210x^4y + 420x^2y^3 - 42y^5 & -42x^5 + 420x^3y^2 - 210xy^4 \end{bmatrix}$$

$$\left\{ -42\left(x^2 + y^2\right)^{\frac{5}{2}} : 1, \ 42\left(x^2 + y^2\right)^{\frac{5}{2}} : 1 \right\}$$

We again obtain two equal eigenvalues with opposite signs. These eigenvalues are zero only at the origin.

9.6.3 *Example: Checking the harmonic bases*

```
1   #Check the rational and logarithmic harmonic bases:
2   x, y = sp.symbols("x, y")
3   X = sp.Matrix([x, y])
4   R2 = x**2 + y**2
5   Rbases = Matrix([x/R2, y/R2, sp.log(R2)])    # put these bases in a list
6
7   nabla2_Rbases = gr.nabla_f(bases, X, dfn=2, smplfy=1)    # Check ∇^2=0?
8   display(Math(f"\\text{{$∇^2$(Rbases)$=0$?}}{latex(nabla2_Rbases.T)}"))
9
10  # Check the hyperbolic harmonic basis functions (HBFs):
11  ω = sp.symbols("ω", real=True)
12  X = sp.Matrix([x, y])
13  HBFs = sp.Matrix([ cos(ω*x)*cosh(ω*y),  cos(ω*y)*cosh(ω*x), #∇^2 HBFs
14                     sin(ω*x)*cosh(ω*y),  cos(ω*y)*sinh(ω*x)])
15  #display(Math(f" \\text{{Harmonic basis functions (HBFs)}}=\
16  #                               {latex(HBFs.T)}"))
17  HBFs = sp.Matrix(HBFs)
18  # Check whether any f satisfies  ∇^2 f = 0:
19  nabla2_HBFs = gr.nabla_f(HBFs, X, dfn=2, smplfy=2)
20  display(Math(f" \\text{{$∇^2$ (HBFs) = 0? }}{latex(nabla2_HBFs.T)}"))
```

$\nabla^2(\text{Rbases}) = 0? \begin{bmatrix} 0 & 0 & 0 \end{bmatrix}$

$\nabla^2 (\text{HBFs}) = 0? \begin{bmatrix} 0 & 0 & 0 & 0 \end{bmatrix}$

9.6.4 *Example: Generating a polynomial CDF field*

We now present an example of creating CDF fields. We first generate some polynomial basis functions using the **gr.create_nbf**() method with different orders and then combine them. In this example, we use a complete third-order set by setting **order = 3**. We then form initial expressions for the potential and stream functions by introducing unknown constants. The code snippets for these tasks are provided as follows:

```
1   # Generate polynomial harmonic bases for potential and stream functions
2   x, y = symbols('x', y')
3   X = sp.Matrix([x, y])
4
5   bfs1 = Matrix([[x], [y]])  # 1st-order harmonic basis (exclude constant)
6   display(Math(f" \\text{{1st-order harmonic bases=}}{latex(bfs1.T)}"))
7   bfsD2 = gr.nabla_f(bfs1, X, dfn=2, smplfy=2)                  # Check ∇**2
8   display(Math(f" \\text{{∇^2 = }}{latex(bfsD2.T)}"))
9
10  order = 3                                     # define the order of bases
11  H_bases = [bfs1]
12  for i in range(2, order+1):          # addition starts from the 2nd order
13      #display(gr.create_nbf(i, X, dfn=2, prnt=False).T)
14      H_bases.append(gr.create_nbf(i, X, dfn=2, prnt=False))     # adds up
15
16  H_bases = Matrix([H_bases])
17  H_bases = H_bases.reshape(H_bases.rows*H_bases.cols, 1)
18  display(Math(f" \\text{{Harmonic bases = }}{latex(H_bases.T)}"))
19
20  ai = sp.symbols(f'a:{len(H_bases)}')             # unknown constants for φ
21  bi = sp.symbols(f'b:{len(H_bases)}')             # unknown constants for ψ
22  ai = Matrix([ai[i] for i in range(len(H_bases))])   # form coeffs vector
23  bi = Matrix([bi[i] for i in range(len(H_bases))])   # form coeffs vector
24
25  cs = ai.col_join(bi)                 # form a single vector of constants
26
27  φg = (ai.T@Matrix(H_bases))[0]              # Form the general functions
28  ψg = (bi.T@Matrix(H_bases))[0]
29
30  display(Math(f" \\text{{Initial potential function, $φ_g$}}:"))
31  display(Math(f" \\text{{ }} {latex(φg)}"))
32  display(Math(f" \\text{{Initial stream function, $ψ_g$}}"))
33  display(Math(f" \\text{{ }} {latex(ψg)}"))
```

First-order harmonic bases $= \begin{bmatrix} x & y \end{bmatrix}$

$\nabla^2 = \begin{bmatrix} 0 & 0 \end{bmatrix}$

Harmonic bases $= \begin{bmatrix} x & xy & x^2y - \frac{y^3}{3} & y & x^2 - y^2 & x^3 - 3xy^2 \end{bmatrix}$

Initial potential function, ϕ_g:

$$a_0 x + a_1 xy + a_2 \left(x^2 y - \frac{y^3}{3} \right) + a_3 y + a_4 \left(x^2 - y^2 \right) + a_5 \left(x^3 - 3xy^2 \right)$$

Initial stream function, ψ_g:

$$b_0 x + b_1 xy + b_2 \left(x^2 y - \frac{y^3}{3} \right) + b_3 y + b_4 \left(x^2 - y^2 \right) + b_5 \left(x^3 - 3xy^2 \right)$$

9.6.5 *General expressions of potential and stream functions*

Each of these initial functions is equipped with six unknown constants, resulting in a total of 12 unknown constants. The next step is to determine these constants. This is accomplished using the Cauchy–Riemann equations (Eq. (9.1)) and the CDF vector component relations (Eq. (9.3)).

Since these equations must be satisfied at any point in the field domain, we need to enforce that all the coefficients of the monomials in these equations equal zero. This leads to a significant number of expressions, some of which may be zero. We will filter out these zero expressions and then solve the nonzero ones. The solution provides the relationships between the unknown constants, which are then used to update the formulas for both the potential and stream functions.

It is also possible — though unlikely — that some expressions do not contain any unknown constants. This implies that these expressions cannot be satisfied regardless of the unknown constants values, and in such cases, we must be vigilant for debugging purposes.

The code snippet for all the tasks mentioned above is provided as follows:

```
 1  ij = [(i, j) for i in range(order + 1)    # for possible monomial orders
 2                      for j in range(order + 2) if i+j <= order-1]
 3  exprs = []
 4  for i in range(len(ij)):                    # use vector component relations:
 5      exprs.append((ϕg.diff(x)-ψg.diff(y)).coeff(x,ij[i][0])\
 6                                              .coeff(y,ij[i][1]))
 7      exprs.append((ϕg.diff(y)+ψg.diff(x)).coeff(x,ij[i][0])\
 8                                              .coeff(y,ij[i][1]))
 9  exprs = gr.remove_zero_exprs(exprs)
10
11  # Find possible expressions that do not contain unknown constants:
12  for i, expr in enumerate(exprs):
13      if len(set(cs) & expr.free_symbols)==0:
14          print(f"Alert: expression cannot be zero: i={i}, expr={expr}")
15
16  display(Math(f" \\text{{Conditions for constants:}}"))
17  display(Math(f" \\text{{ }}{latex(exprs)}"))
18
19  sln_ab = sp.solve(exprs, cs)
20  display(Math(f" \\text{{Constant relations found:}}\\;{latex(sln_ab)}"))
21
22  display(Math(f" \\text{{General potential function, $ϕ_g$:}}"))
23  display(Math(f" \\;{latex(ϕg.subs(sln_ab))}"))
24  display(Math(f" \\text{{General  stream  function, $ψ_g$}}"))
25  display(Math(f" \\;\\; {latex(ψg.subs(sln_ab))}"))
```

```
Total number of expressions = 12
The Index of the zero expressions: [4, 5, 10, 11]
Number of nonzero expressions = 8
```

Conditions for constants:

$$\begin{bmatrix} a_0 - b_3, \ a_3 + b_0, \ a_1 + 2b_4, \ -2a_4 + b_1, \ 2a_4 - b_1, \ a_1 + 2b_4, \\ 2a_2 + 6b_5, \ -6a_5 + 2b_2 \end{bmatrix}$$

Constant relations found:

$$\left\{ a_0 : b_3, \ a_1 : -2b_4, \ a_2 : -3b_5, \ a_3 : -b_0, \ a_4 : \frac{b_1}{2}, \ a_5 : \frac{b_2}{3} \right\}$$

General potential function, ϕ_g:

$$-b_0 y + \frac{b_1 \left(x^2 - y^2 \right)}{2} + \frac{b_2 \left(x^3 - 3xy^2 \right)}{3} + b_3 x - 2b_4 xy - 3b_5 \left(x^2 y - \frac{y^3}{3} \right)$$

General stream function, ψ_g:

$$b_0 x + b_1 xy + b_2 \left(x^2 y - \frac{y^3}{3} \right) + b_3 y + b_4 \left(x^2 - y^2 \right) + b_5 \left(x^3 - 3xy^2 \right)$$

9.6.6 *Potential and stream functions for particular CDF fields*

The general potential and stream functions satisfy all the necessary conditions for a CDF field. However, they still contain constants, implying that any set of specified constants can be used to produce a CDF field. Thus, we have a general expression for all (infinitely many) possible polynomial CDF fields for orders below 3. For example:

1. Setting $b_0 = gm$ and all other $b_i = 0$ yields the constant gravity field discussed in Section 9.3.4.
2. Setting $b_1 = -1$ and all other $b_i = 0$ results in the linear field covered in Section 9.3.3.

Readers may experiment with this by uncommenting the corresponding code lines here. In the following example, we set $b_2 = -1$, $b_5 = 1$, and all other $b_i = 0$. This generates a particular quadratic CDF field using the snippet provided as follows:

```
1   # The gravity CDF field: [0  -g*m]^T :
2   #dic_ab = {bi[0]:g*m,bi[1]:0,bi[2]:0,bi[3]:0,bi[4]:0,bi[5]:0}
3
4   # The linear CDF field: [-x  y]^T :
5   #dic_ab = {bi[0]:0,bi[1]:-1,bi[2]:0,bi[3]:0,bi[4]:0,bi[5]:0}
6
7   # Parameters for a quadratic CDF field (this example):
8   dic_ab = {bi[0]:0,bi[1]:0,bi[2]:-1,bi[3]:0,bi[4]:0,bi[5]:1}
9
10  ϕ = ϕg.subs(sln_ab).subs(dic_ab)
11  ψ = ψg.subs(sln_ab).subs(dic_ab)
12  display(Math(f" \\text{{Potential function = }} {latex(ϕ)}; \\;\\;\\;\
13                  \\text{{Stream function = }} {latex(ψ)}"))
14
15  Vx = ψ.diff(y); Vy = -ψ.diff(x);
16  display(Math(f" \\text{{Flow field $V_{{x}}$ = }}{latex(Vx)}; \\;\\;\\;\
17                  \\text{{ $V_{{y}}$ = }}{latex(Vy)}"))
18  orth_ϕψ = gr.grad_f(ϕ,X).T@gr.grad_f(ψ,X)
19  display(Math(f" \\text{{Orthogonal?}}\\;\
20                  {latex(orth_ϕψ[0].simplify()==0)}"))
21
22  Vx = ϕ.diff(x); Vy = ϕ.diff(y)
23  display(Math(f" \\text{{Flow field via ϕ, $V_{{x}}$ = }}{latex(Vx)};\
24                  \\;\\;\\; \\text{{ $V_{{y}}$ = }}{latex(Vy)}"))
25  Vx = ψ.diff(y); Vy = -ψ.diff(x)
26  display(Math(f" \\text{{Flow field via ψ, $V_{{x}}$ = }}{latex(Vx)};\
27                  \\;\\;\\; \\text{{ $V_{{y}}$ = }}{latex(Vy)}"))
```

Potential function $= -\dfrac{x^3}{3} - 3x^2y + xy^2 + y^3$;

Stream function $= x^3 - x^2y - 3xy^2 + \dfrac{y^3}{3}$

Flow field $V_x = -x^2 - 6xy + y^2$; $\quad V_y = -3x^2 + 2xy + 3y^2$

Orthogonal? True

Flow field via ϕ, $V_x = -x^2 - 6xy + y^2$; $\quad V_y = -3x^2 + 2xy + 3y^2$

Flow field via ψ, $V_x = -x^2 - 6xy + y^2$; $\quad V_y = -3x^2 + 2xy + 3y^2$

9.6.7 *Plot the created quadratic CDF field*

Finally, we use plot_potentialStream_f() to plot the quadratic CDF field created above:

```
1  # Plot the vector field:
2  plt.ioff()
3  #plt.ion()                        # use plt.ion() to generate a new plot
4  C = np.arange(-2, 2.0, 0.2)       # constants for potential & stream f
5  xL = -2.0; xR = 2.0
6  yL = 0.01; yU = 2.0
7  plot_potentialStream_f(φ, ψ, C, xL, xR, yL, yU, contourδ=0.0005)
8
9  # Two saddle points:
10 plt.scatter(0, 0, color='r', s=30, zorder=3)
11
12 plt.savefig('imagesVC/potentialStream_3rd.png', dpi=500) #save to file
13 #plt.show()
```

We have discovered a new polynomial potential incompressible vector field. Both the potential and stream functions for this vector field are illustrated in Fig. 9.3. The key features of this field are as follows:

1. The contour lines of the potential and stream functions are orthogonal to each other at every point in the domain.
2. Every point within the domain of the CDF vector function is a saddle point, with fluid flowing into the point from one direction and out from it in another direction.
3. These functions are quadratic in nature.

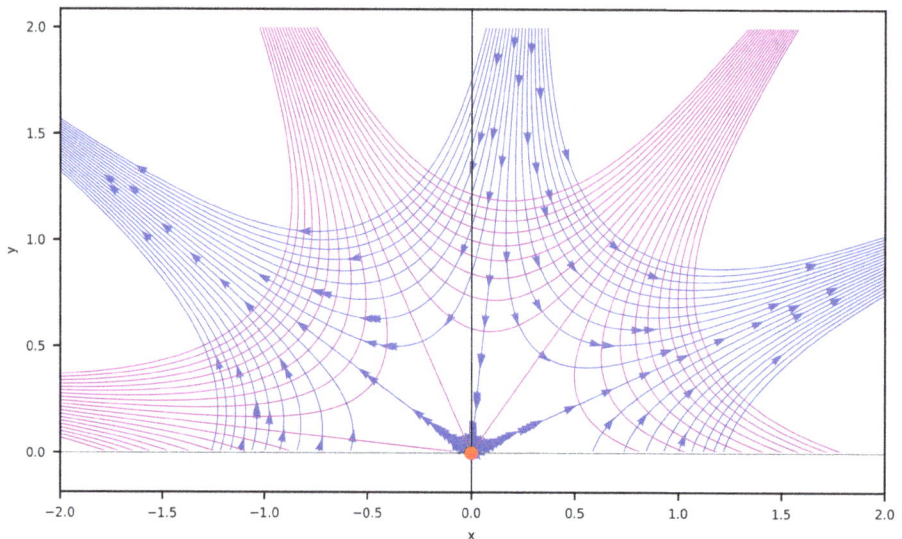

Figure 9.3. Schematic view of the potential function (magenta) and stream function (blue) for the quadratic field created using third-order harmonic polynomial basis functions.

Next, let us examine the eigenvalues of the Jacobian matrix for this potential flow field:

```
1  J_3rd = sp.Matrix([[Vx.diff(x), Vx.diff(y)],
2                     [Vx.diff(y),-Vx.diff(x)]])
3  display(Math(f" \\text{{Jacobian matrix using 3rd-order bases = }}\
4                            {latex(J_3rd)}"))
5  J_3rd.eigenvals()
```

Jacobian matrix using third-order bases $= \begin{bmatrix} -2x - 6y & -6x + 2y \\ -6x + 2y & 2x + 6y \end{bmatrix}$

$$\left\{ -2\sqrt{10}\sqrt{x^2 + y^2} : 1, \ 2\sqrt{10}\sqrt{x^2 + y^2} : 1 \right\}$$

We again obtain two equal eigenvalues with opposite signs. These eigenvalues are zero only at the origin.

Readers are encouraged to generate additional CDF fields using the polynomial bases and explore interesting configurations.

Furthermore, we invite readers to experiment with other bases, such as hyperbolic harmonic bases, to create new CDF fields. A comparison can then be made with the field generated using the complex function $\sin(\omega z)$.

9.7 A CDF field via rational harmonic basis functions

```
1  # Generate polynomial harmonic bases for potential and stream functions
2  x, y = symbols('x, y')
3  X = sp.Matrix([x, y])
4
5  RBFs = Matrix([x/(x**2 + y**2), y/(x**2 + y**2)])        # Rational bases
6  #RBFs = Matrix([sp.log(x**2 + y**2)])                    # Logarithmic bases
7
8  display(Math(f" \\text{{Rational bases=}}{latex(RBFs.T)}"))
9  nabla2_RBFs = gr.nabla_f(RBFs, X, dfn=2, smplfy=1)       # Check ∇**2
10 display(Math(f" \\text{{∇^2 = }}{latex(nabla2_RBFs.T)}"))
11
12 ai = sp.symbols(f'a:{len(RBFs)}')          # unknown constants for φ
13 bi = sp.symbols(f'b:{len(RBFs)}')          # unknown constants for ψ
14 ai = Matrix([ai[i] for i in range(len(RBFs))])    # form coeffs vector
15 bi = Matrix([bi[i] for i in range(len(RBFs))])    # form coeffs vector
16
17 cs = ai.col_join(bi)                       # form a single vector of constants
18
19 φg = (ai.T@Matrix(RBFs))[0]                # Form the general functions
20 ψg = (bi.T@Matrix(RBFs))[0]
21
22 display(Math(f" \\text{{Initial potential function, $φ_g$}}=\
23                            {latex(φg)}"))
24 display(Math(f" \\text{{Initial stream function, $ψ_g$}}={latex(ψg)}"))
```

$$\text{Rational bases} = \begin{bmatrix} \frac{x}{x^2+y^2} & \frac{y}{x^2+y^2} \end{bmatrix}$$

$$\nabla^2 = \begin{bmatrix} 0 & 0 \end{bmatrix}$$

$$\text{Initial potential function, } \phi_g = \frac{a_0 x}{x^2 + y^2} + \frac{a_1 y}{x^2 + y^2}$$

$$\text{Initial stream function, } \psi_g = \frac{b_0 x}{x^2 + y^2} + \frac{b_1 y}{x^2 + y^2}$$

9.7.1 *General expressions of potential and stream functions*

It can be observed that each of these initial functions is associated with two unknown constants, resulting in a total of four unknown constants that need to be determined. This process is carried out using the Cauchy–Riemann equation (9.1) and the CDF vector component relations equation (9.3) similar to the previous example.

Since these functions are rational, we must simplify them to ensure that the expressions have common denominators. Once this is achieved, we can enforce that all the coefficients of the monomials in the numerators are equal to zero. The code snippet for executing all the aforementioned tasks is as follows:

```
1  eqn1 = (ϕg.diff(x)-ψg.diff(y)).simplify()
2  display(Math(f"\\text{{An example of rational equation=}}{latex(eqn1)}"))
3  numer_eqn1 = sp.numer((ϕg.diff(x)-ψg.diff(y)).simplify()).expand()
4  display(Math(f"\\text{{ Numerator (must be 0) =}}{latex(numer_eqn1)}"))
```

An example of rational equation

$$= \frac{-2a_0 x^2 - 2a_1 xy + 2b_0 xy + 2b_1 y^2 + (a_0 - b_1)\left(x^2 + y^2\right)}{\left(x^2 + y^2\right)^2}$$

Numerator (must be 0) $= -a_0 x^2 + a_0 y^2 - 2a_1 xy + 2b_0 xy - b_1 x^2 + b_1 y^2$

```
1  # Find the expressions for these constants:
2
3  ij = [(i, j) for i in range(order + 1)    # for possible monomial orders
4                        for j in range(order + 2) if i+j <= order-1]
5
6  # using the components relations:
7  numer_eqn1 = sp.numer((ϕg.diff(x)-ψg.diff(y)).simplify()).expand()
8  numer_eqn2 = sp.numer((ϕg.diff(y)+ψg.diff(x)).simplify()).expand()
9
10 exprs = []
11 for i in range(len(ij)):               # use vector component relations:
12     exprs.append(numer_eqn1.coeff(x,ij[i][0]).coeff(y,ij[i][1]))
13     exprs.append(numer_eqn1.coeff(x,ij[i][0]).coeff(y,ij[i][1]))
14 exprs = gr.remove_zero_exprs(exprs)
15
16 # Find possible expressions that do not contain unknown constants:
17 for i, expr in enumerate(exprs):
18     if len(set(cs) & expr.free_symbols)==0:
19         print(f"Alert: expression cannot be zero: i={i}, expr={expr}")
20
21 display(Math(f" \\text{{Conditions for constants:}}"))
22 display(Math(f" \\text{{ }}{latex(exprs)}"))
23
24 sln_ab = sp.solve(exprs, cs)
25 display(Math(f" \\text{{Constant relations found:}}\\;{latex(sln_ab)}"))
26
27 display(Math(f" \\text{{General potential function, $ϕ_g$ = }} \
28                       {latex(ϕg.subs(sln_ab))}"))
29 display(Math(f" \\text{{General  stream   function, $\;\;\, ψ_g$ = }}\
30                       {latex(ψg.subs(sln_ab))}"))
```

```
Total number of expressions = 12
The Index of the zero expressions: [0, 1, 2, 3, 6, 7]
Number of nonzero expressions = 6
```

Conditions for constants:

$$[a_0 + b_1, \ a_0 + b_1, \ -2a_1 + 2b_0, \ -2a_1 + 2b_0, \ -a_0 - b_1, \ -a_0 - b_1]$$

Constant relations found: $\{a_0 : -b_1, \ a_1 : b_0\}$

General potential function, $\phi_g = \dfrac{b_0 y}{x^2 + y^2} - \dfrac{b_1 x}{x^2 + y^2}$

General stream function, $\psi_g = \dfrac{b_0 x}{x^2 + y^2} + \dfrac{b_1 y}{x^2 + y^2}$

9.7.2 *Potential and stream functions for a particular CDF field*

The general potential and stream functions derived previously satisfy all the necessary conditions for a CDF field; however, they still contain two constants. This implies that any chosen set of constants can be utilized to produce a specific CDF field. Consequently, we have a general expression for an infinite number of possible polynomial CDF fields. In this example, let us set $b_0 = b_1 = 1$. A particular CDF field can then be generated using the following code snippet:

```
1  # Parameters for a quadratic CDF field (this example):
2  dic_ab = {bi[0]:1,bi[1]:1}
3
4  ϕ = ϕg.subs(sln_ab).subs(dic_ab)
5  ψ = ψg.subs(sln_ab).subs(dic_ab)
6  display(Math(f" \\text{{Potential function = }} {latex(ϕ)}; \\;\\;\\;\
7                \\text{{Stream function = }} {latex(ψ)}"))
8
9  orth_ϕψ = gr.grad_f(ϕ,X).T@gr.grad_f(ψ,X)
10 display(Math(f" \\text{{Orthogonal?}}\\;\
11                {latex(orth_ϕψ[0].simplify()==0)}"))
12
13 Vx = ϕ.diff(x).expand().simplify()
14 Vy = ϕ.diff(y).expand().simplify()
15 display(Math(f" \\text{{Flow field via ϕ, $V_{{x}}$ = }}{latex(Vx)};\
16                \\;\\;\\; \\text{{ $V_{{y}}$ = }}{latex(Vy)}"))
17 Vx = ψ.diff(y).expand().simplify()
18 Vy =-ψ.diff(x).expand().simplify()
19 display(Math(f" \\text{{Flow field via ψ, $V_{{x}}$ = }}{latex(Vx)};\
20                \\;\\;\\; \\text{{ $V_{{y}}$ = }}{latex(Vy)}"))
```

Potential function $= -\dfrac{x}{x^2+y^2} + \dfrac{y}{x^2+y^2};$

Stream function $= \dfrac{x}{x^2+y^2} + \dfrac{y}{x^2+y^2}$

Orthogonal? True

Flow field via ϕ, $V_x = \dfrac{x^2 - 2xy - y^2}{x^4 + 2x^2y^2 + y^4};$ $V_y = \dfrac{x^2 + 2xy - y^2}{x^4 + 2x^2y^2 + y^4}$

Flow field via ψ, $V_x = \dfrac{x^2 - 2xy - y^2}{x^4 + 2x^2y^2 + y^4};$ $V_y = -\dfrac{-x^2 - 2xy + y^2}{x^4 + 2x^2y^2 + y^4}$

```
1  # The eigenvalues of the Jacobian matrix:
2  J_Rational = sp.Matrix([[Vx.diff(x).simplify(), Vx.diff(y).simplify()],
3                          [Vx.diff(y).simplify(),-Vx.diff(x).simplify()]])
4  display(Math(f" \\text{{Jacobian matrix using Rational bases = }}\
5                {latex(J_Rational)}"))
6  J_Rational.eigenvals()
```

Jacobian matrix using rational bases

$$= \begin{bmatrix} \dfrac{2(-x^3+3x^2y+3xy^2-y^3)}{x^6+3x^4y^2+3x^2y^4+y^6} & \dfrac{2(-x^3-3x^2y+3xy^2+y^3)}{x^6+3x^4y^2+3x^2y^4+y^6} \\ \dfrac{2(-x^3-3x^2y+3xy^2+y^3)}{x^6+3x^4y^2+3x^2y^4+y^6} & -\dfrac{2(-x^3+3x^2y+3xy^2-y^3)}{x^6+3x^4y^2+3x^2y^4+y^6} \end{bmatrix}$$

$$\left\{ -2\sqrt{2}\sqrt{\dfrac{1}{(x^2+y^2)^3}} : 1,\ 2\sqrt{2}\sqrt{\dfrac{1}{(x^2+y^2)^3}} : 1 \right\}$$

We obtain again two equal eigenvalues with opposite signs. This CDF field has a singularity at the origin.

9.7.3 *Plotting the created rational CDF field*

Finally, we use the `plot_potentialStream_f()` function to plot the CDF field created above:

```
1  # Plot the vector field:
2  plt.ioff()
3  #plt.ion()                          # use plt.ion() to generate a new plot
4  C = np.arange(-2, 2.0, 0.15)        # constants for potential & stream f
5  xL = -2.6; xR = 2.6 # xL = -2.0; xR = 2.0
6  yL = -1.5; yU = 1.5 # yL = 0.01; yU = 2.0
7  plot_potentialStream_f(ϕ, ψ, C, xL, xR, yL, yU, contourδ=0.0005)
8
9  plt.savefig('imagesVC/potentialStream_Rational.png', dpi=500)
10 #plt.show()
```

We have identified yet another new CDF field. Both the potential and stream functions for this vector field are plotted in Fig. 9.4. It exhibits the following features:

1. The contour lines of the potential and stream functions are orthogonal to each other at any point in the domain, which is typical for any CDF field.

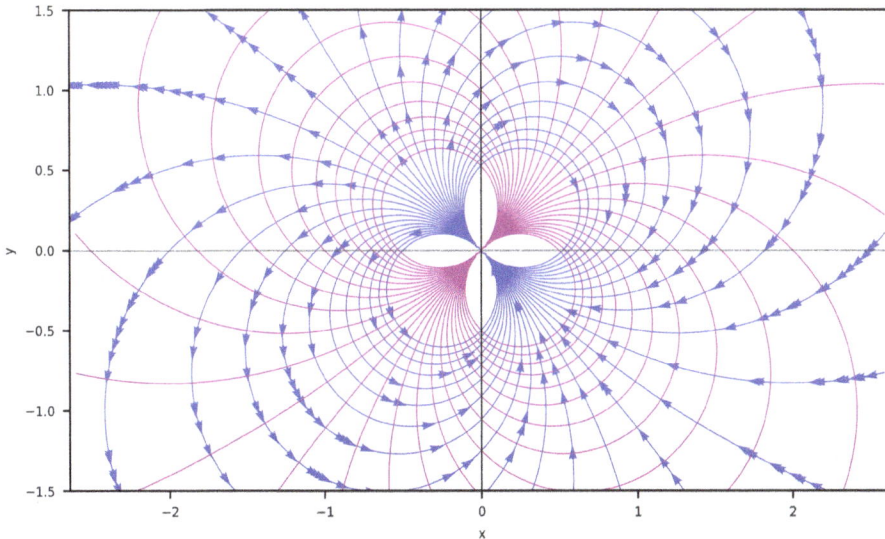

Figure 9.4. Schematic view of the potential function (magenta) and stream function (blue) for the rational field.

2. Every point in the domain of the CDF vector function is a saddle point except for the singular point at the origin. The fluid flows into these saddle points in one direction and out from them in another direction.
3. The singular point both emits and swallows fluid.

9.8 Logarithmic harmonic functions

In Section 9.6.1, a pair of logarithmic harmonic functions is listed. One is a real function, and the other is a complex function. The real function is well-known and can be found in the literature. The complex function is derived in this work as the stream function by assuming that the vector field generated by the real potential function is a CDF field. The code snippet is as follows:

```
1   # Generate polynomial harmonic bases for potential and stream functions
2   x, y = symbols('x, y')
3   X = sp.Matrix([x, y])
4
5   LBFs = sp.log(x**2 + y**2)                              # Logarithmic basis
6
7   display(Math(f" \\text{{Rational bases=}}{latex(LBFs)}"))
8   nabla2_LBFs = gr.nabla_f(LBFs, X, dfn=2, smplfy=1)         # Check ∇**2
9   display(Math(f" \\text{{$∇^2$(LBFs)= }}{latex(nabla2_LBFs)}"))
10
11  a = sp.symbols('a')                       # unknown constants for φ
12  φg = a*LBFs                                # Form the general functions
13
14  display(Math(f" \\text{{Initial potential function, $φ_g$}}=\
15                              {latex(φg)}"))
16  Vx = φg.diff(x).expand().simplify()
17  Vy = φg.diff(y).expand().simplify()
18  display(Math(f" \\text{{Flow field via φ, $V_{{x}}$ = }}{latex(Vx)};\
19                      \\;\\;\\; \\text{{ $V_{{y}}$ = }}{latex(Vy)}"))
20
21  ψ = gr.find_stream_f2D(Vx, Vy, X)          # Assume [Vx  Vy] a CDF field
22  display(Math(f" \\text{{The stream function ψ found}}={latex(ψ)}"))
23  nabla2_ψ = gr.nabla_f(ψ, X, dfn=2, smplfy=1)          # Check ∇^2=0?
24  display(Math(f" \\text{{Does ψ satisfy $∇^2ψ$=0?}}=\
25                              {latex(nabla2_ψ==0)}"))
26  Vx = ψ.diff(y).simplify(); Vy =-ψ.diff(x).simplify()      # vector field
27  display(Math(f" \\text{{Vector field found via ψ}}={latex((Vx, Vy))}"))
```

Rational bases$= \log \left(x^2 + y^2 \right)$

$\nabla^2 (\text{LBFs}) = 0$

Initial potential function, $\phi_g = a \log \left(x^2 + y^2 \right)$

Flow field via ϕ, $V_x = \dfrac{2ax}{x^2 + y^2}; \quad V_y = \dfrac{2ay}{x^2 + y^2}$

```
ψ = I*a*(log(x - I*y) - log(x + I*y))
(ψ.diff(x)+Vy).simplify() = 0
(ψ.diff(y)-Vx).simplify() = 0
```

The stream function ψ found $= ia\left(\log\left(x - iy\right) - \log\left(x + iy\right)\right)$

Does ψ satisfy $\nabla^2\psi=0$? True

Vector field found via $\psi = \left(\dfrac{2ax}{x^2 + y^2}, \dfrac{2ay}{x^2 + y^2}\right)$

We have derived the general expressions for both the potential and stream functions for this CDF field, which contains only one constant a. By setting a to a specific value, we can produce a particular CDF field.

The significant difference between this CDF field and the others discussed earlier is that the stream function is a complex function. Consequently, the streamlines do not reside in real space but in the extended complex space, making it challenging to plot the potential and stream functions together.

Readers may compute the eigenvalues of the Jacobian matrix for this field.

9.9 Concluding remarks

This chapter has focused on a special type of vector field: CDF fields. We have examined the unique features of CDF fields and the presented techniques, along with code, to construct them. We conclude the chapter with the following remarks:

1. A CDF field is conservative, irrotational, and divergence-free. It possesses both a potential and a stream function.
2. The potential and stream functions are orthogonal to each other at every point in the domain, which is a characteristic feature of CDF fields.
3. Every point in the domain of the CDF vector function is a saddle point except for singular points. The fluid flows into these saddle points from one direction and out from another.
4. There exist points that emit fluid and points that swallow fluid.
5. The Cauchy–Riemann equations and vector component relations are utilized to generate CDF fields supported by various harmonic basis functions.
6. A number of interesting CDF fields have been discovered and plotted.

The following chapter will discuss the techniques for approximating the derivatives of functions using the theorems of vector fields.

References

[1]   G. Strang, *Calculus*, Wellesley-Cambridge Press, Massachusetts, 1991.
[2]   G.R. Liu, *Numbers and Functions: Theory, Formulation, and Python Codes*, World Scientific, New Jersey, 2024.
[3]   G.R. Liu, *Mechanics of Materials: Formulations and Solutions with Python*, World Scientific, New Jersey, 2024.
[4]   G.R. Liu, *Calculus: A Practical Course with Python*, World Scientific, New Jersey, 2025.
[5]   G.R. Liu, *Solid Mechanics with Python: In Cartesian Coordinates*, World Scientific, New Jersey, 2024.

Chapter 10

Gradient Smoothing Methods

```
 1  # Place cursor in this cell, and press Ctrl+Enter to import dependences.
 2  import sys                       # For accessing the computer system
 3  sys.path.append('../grbin/')        # Add in the path to your system
 4
 5  from commonImports import *      # Import dependences from '../grbin/'
 6  import grcodes as gr              # Import the module of the author
 7  importlib.reload(gr)             # When grcodes is modified, reload it
 8
 9  init_printing(use_unicode=True)       # For latex-like quality printing
10  np.set_printoptions(precision=4,suppress=True,
11                      formatter={'float_kind': '{:.4e}'.format})
```

In previous chapters, we studied several theorems, including Green's theorem, divergence theorem, Stokes' theorem, Gauss's formula, and formulas extended from Gauss's formula. We also discussed various applications of these formulas. This chapter focuses on one of their applications: computing smoothed derivatives, curl, divergence, or gradient. This technique is known as the gradient smoothing method (GSM) [1–3], as it leverages the gradient of a scalar function, encompassing all partial derivatives. GSM computes the partial derivatives of a function without differential operations, instead utilizing integrations along the boundary of a smoothing domain. This approach provides an accurate and stable technique for approximating the partial derivatives of functions, especially when the function is approximated and may not even be differentiable. GSM has broad applications in computational methods.

This chapter was developed with reference to textbooks in Refs. [4–6]. Wikipedia pages, particularly those on integral (https://en.wikipedia .org/wiki/Integral), surface integral (https://en.wikipedia.org/wiki/Surface_ integral), and Green's theorem (https://en.wikipedia.org/wiki/Green%27s_

theorem), also serve as valuable additional resources. Both NumPy and SymPy are used in developing the code for the demonstration examples. Discussions with ChatGPT, Gemini, and Bing were highly beneficial for coding and for preparing this chapter.

10.1 Computation of the smoothed gradient of scalar fields

As mentioned earlier, the functions M and N do not have any relation; the only condition is that they must be differentiable. Using Eqs. (4.2) and (4.3), and setting both M and N as the generic function $f(x, y)$ in a 2D space, we have

$$
\frac{1}{A} \oint_C f \, \mathrm{d}y = \underbrace{\frac{1}{A} \iint_D \left(\frac{\partial f}{\partial x} \right) \mathrm{d}A}_{\text{smoothed partial derivative of } f \text{ wrt } x, \ \overline{\frac{\partial f}{\partial x}}}
$$

$$
-\frac{1}{A} \oint_C f \, \mathrm{d}x = \underbrace{\frac{1}{A} \iint_D \left(\frac{\partial f}{\partial y} \right) \mathrm{d}A}_{\text{smoothed partial derivative of } f \text{ wrt } y, \ \overline{\frac{\partial f}{\partial y}}}
$$

(10.1)

where A is the area of domain $\mathcal{D}$. Equation (10.1) can be written as

$$
\overline{\nabla f} = \underbrace{\begin{bmatrix} \overline{\frac{\partial f}{\partial x}} \\ \overline{\frac{\partial f}{\partial y}} \end{bmatrix}}_{\text{smoothed gradient, } \overline{\nabla f}} = \begin{bmatrix} \frac{1}{A} \oint_C f \mathrm{d}y \\ -\frac{1}{A} \oint_C f \mathrm{d}x \end{bmatrix}
$$

(10.2)

We use an overhead bar to represent a smoothed quantity. It can be observed that the smoothed gradient or partial derivatives of the function f can be computed using only the values of f on the boundary curve through curve integration (not differentiation).

Note also that the surface $\mathcal{C}$ in Eq. (10.2) can have an arbitrary shape and size, and can be located anywhere in the domain. The only requirement for $\mathcal{C}$ is that it be at least piecewise smooth. The domain $\mathcal{D}$ enclosed by the curve $\mathcal{C}$ is often referred to as the **smoothing domain**.

10.1.1 *Case study: Smoothed gradients of functions*

Consider scalar functions that are polynomials defined in 2D as

$$f_1(x, y, z) = x + 2y + 8; \quad \text{linear}$$
$$f_2(x, y, z) = 11x^2 + 22y^2 + x + 2y + 8; \quad \text{quadratic} \qquad (10.3)$$
$$f_3(x, y, z) = 11x^3 + 22y^2 + x + 2y + 8; \quad \text{third order}$$

The gradients of these functions can be computed easily through differentiation. However, we aim to compute their smoothed gradients without using differentiation. We consider two different smoothing domains: a circle, shown in Fig. 10.1(a), and a polygon, shown in Fig. 10.1(b).

The tasks for this study are as follows:

1. Compute the exact gradient of the given functions.
2. Using the circular smoothing domain shown in Fig. 10.1(a), compute the smoothed gradient of the linear function $f_1(x, y)$ at the center of the circle.
3. Repeat item 2 for the quadratic function $f_2(x, y)$.
4. Repeat item 2 for the cubic (third-order) function $f_3(x, y)$.
5. Repeat items 2–4 using a pentagon, shown in Fig. 10.1(b), as the smoothing domain.
6. Discuss the results obtained.

Solution to Question 1: This is straightforward as it involves only partial differentiation of the function. We have developed the Python function `gr.grad_f()` for this purpose in Ref. [7]. The following code snippet computes the exact gradients of these three functions. The results will be used for comparison with the smoothed gradients to be obtained later:

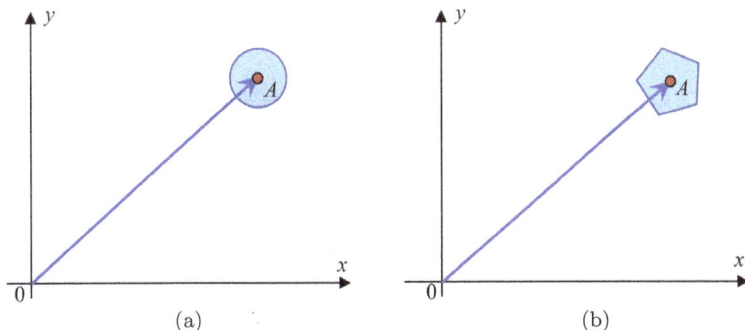

(a)   (b)

Figure 10.1. (a) A circle and (b) a polygon domain used as smoothing domains for computing the smoothed gradient of a scalar function at the center of the domains.

```
1  # Exact solution for the gradient of the function:
2
3  x, y = symbols('x, y ', real=True)
4  X = Matrix([x, y ])
5
6  f1 = 1*x + 2*y + 8                                            # linear
7  f2 = 11*x**2 + 22*y**2 + 1*x+ 2*y + 8                        # quadratic
8  f3 = 11*x**3 + 22*y**2 + 1*x+ 2*y + 8                          # cubic
9
10  display(Math(f" \\text{{Exact gradient of these functions:}}"))
11  display(Math(f" \\text{{∇}}({latex(f1)}) = {latex(gr.grad_f(f1, X))}"))
12  display(Math(f" \\text{{∇}}({latex(f2)}) = {latex(gr.grad_f(f2, X))}"))
13  display(Math(f" \\text{{∇}}({latex(f3)}) = {latex(gr.grad_f(f3, X))}"))
```

Exact gradient of these functions:

$$\nabla(x + 2y + 8) = \begin{bmatrix} 1 \\ 2 \end{bmatrix}$$

$$\nabla(11x^2 + x + 22y^2 + 2y + 8) = \begin{bmatrix} 22x + 1 \\ 44y + 2 \end{bmatrix}$$

$$\nabla(11x^3 + x + 22y^2 + 2y + 8) = \begin{bmatrix} 33x^2 + 1 \\ 44y + 2 \end{bmatrix}$$

As expected, the gradient of the linear function is a constant vector function; the gradient of the quadratic function is a linear vector function, and the gradient of the cubic function is a quadratic vector function.

To address the other questions, we provide the Python function in the following section, which can be used multiple times in our study.

10.1.2 *Code for the smoothed gradient over a circle*

The following Python function computes the smoothed gradient of a given scalar function $f(x, y)$ over a circular domain with a radius of a. The description and tasks for the code are listed in the docstring:

```
1  def gsm_circle(f, X, center, a):
2      '''
3      Compute the smoothed gradient of a scalar function f using a
4      circular curve with radius of r=a. It prints out
5          1 The gradient of f.
6          2 Smoothed gradient of f.
7          3 center: the center of the circle.
8      '''
```

```
9    θ = symbols('θ', real=True)          # polar angle as the parameter
10   x, y = X                              # unpack the Cartesian coordinates
11   A = sp.pi*a**2                        # area of the circle
12
13   dicX0 = {x:center[0], y:center[1]}
14   gradf = gr.grad_f(f, X).subs(dicX0)          # exact gradient of f
15   display(Math(f" \\text{{Exact gradient of f = }}{latex(gradf)}"))
16
17   # Coordinate transformation, for integral over the closed circle C:
18   xC = a*cos(θ) + center[0]             # xC, yC are x,y on curve C
19   yC = a*sin(θ) + center[1]
20   dicC = {x:xC, y:yC}
21
22   fIC = f.subs(dicC)
23   #display(Math(f"\\text{{ Integrand of the curve integral}}=\
24   #                    {latex(sp.simplify(fIC))}"))
25   dx = xC.diff(θ); dy = yC.diff(θ)
26   Int = integrate
27   s_dfdx =  (Int(fIC*dy, (θ,0,2*sp.pi))/A).simplify() #smoothed diffs
28   s_dfdy = -(Int(fIC*dx, (θ,0,2*sp.pi))/A).simplify()
29   s_gradf = Matrix([s_dfdx, s_dfdy])            # smoothed gradient
30
31   display(Math(f" \\text{{Smoothed gradient of f}}=\
32                       {latex(s_gradf)}"))
```

Solution to Question 2: We use gsm_circle() to compute the smoothed gradient of the linear function:

```
1  a, c = symbols('a, c', nonnegative=True)      # circle radius and center
2
3  center = Matrix([2*c, 5*c])        # place the center of the sphere here
4  gsm_circle(f1, X, center, a)
```

Exact gradient of f $= \begin{bmatrix} 1 \\ 2 \end{bmatrix}$

Smoothed gradient of f $= \begin{bmatrix} 1 \\ 2 \end{bmatrix}$

It is found that the smoothed gradient of the linear function matches the exact gradient regardless of the size of the smoothing domain and its location.

Solution to Question 3: Compute the smoothed gradient of the quadratic function using the same Python function:

```
1  gsm_circle(f2, X, center, a)
```

Exact gradient of f $= \begin{bmatrix} 44c + 1 \\ 220c + 2 \end{bmatrix}$

Smoothed gradient of f $= \begin{bmatrix} 44c + 1 \\ 220c + 2 \end{bmatrix}$

In the results provided, c controls the location of the circle, while a controls its size. It is observed that the smoothed gradient of the quadratic function is independent of the circle's size a but depends on the center's location c. This is logical as changing the location affects the function's gradient. However, the smoothed gradient remains consistent with the exact gradient at the center of the smoothing domain. This holds true regardless of the size and location of the smoothing domain, indicating that Eq. (10.2) yields the exact gradient for both linear and quadratic scalar functions when employing a circular smoothing domain.

Solution to Question 4: Compute the smoothed gradient of the third-order function using the same Python function:

```
1  gsm_circle(f3, X, center, a)  # one my try to replace f3 with x**5+y**5
```

Exact gradient of f $= \begin{bmatrix} 132c^2 + 1 \\ 220c + 2 \end{bmatrix}$

Smoothed gradient of f $= \begin{bmatrix} \frac{33a^2}{4} + 132c^2 + 1 \\ 220c + 2 \end{bmatrix}$

This time, we found that the smoothed gradient differs from the exact one. The smoothed gradient depends on the radius of the circle a. As we shrink the circle, the two gradients converge and become identical when $a = 0$. This indicates that Eq. (10.2) serves as an approximation of the exact gradient for functions of order higher than 2.

It's important to note that the error's dependence on the smoothing domain dimension a is quadratic, reflecting the third-order nature of the function. For a fourth-order function, the dependence could be cubic in a, and for a fifth-order function, it could be quartic, and so forth. You may explore this further using the code provided above.

Additionally, the GSM can produce the exact gradient by employing the boundary integral of a smoothing domain and setting the size of the smoothing domain to zero.

Solution to Question 5: To compute the smoothed gradient of these functions over a polygonal domain, we write the Python function in the following section. Here, we utilize techniques for curve integration along straight lines and for calculating the area of a polygon with an arbitrary number of edges using polygon_area_green().

10.1.3 *Code for the smoothed gradient over a polygon*

The following Python function computes the smoothed gradient of a given scalar function $f(x, y)$ over a general polygonal domain consisting solely of straight lines. The tasks performed by the code are detailed in the docstring of the function:

```python
 1  def gsm_polygon(f, X, nodes, p_out=True):
 2      '''Computes the smoothed gradient of a given scalar function f
 3      (Sympy) over a general polygonal domain defined by its nodes.
 4      Example: nodes = np.array([[-1,-1], [ 1,-1], [ 1, 1], [-1, 1]]).
 5      return: s_gradf - smoothed gradient of f.
 6             gradf - exact gradient of f.
 7      '''
 8      x, y = X
 9      t = sp.symbols('t')                         # parameter for line integration
10
11      #A = gr.polygon_area(nodes)        # area of the polygon, alternative
12      A = gr.polygon_area_green(nodes)
13      gradf = gr.grad_f(f, X).subs(dict(zip(X, np.mean(nodes, axis=0))))
14      gradf = sp.simplify(gradf.expand()).evalf(6)
15
16      # We roll the nodes back by 1, and use it to form the polygon edges.
17      nodes_r1 = np.roll(nodes, -1, axis=0)
18      r = nodes + t*(nodes_r1-nodes)  # parameterization with t for edges
19      drdt = sp.diff(sp.Matrix(r), t)     # drdt[i,0]=dxdt; drdt[i,1]=dydt
20      t_limits = (t, 0, 1)                 # limits for t (0-1 for each edge)
21
22      Int = sp.integrate
23      s_dfdx = 0; s_dfdy = 0
24      for i in range(len(nodes)): # line integrals over dt along each edge
25          s_dfdx += Int((f*drdt[i,1]).subs({x:r[i,0], y:r[i,1]}),t_limits)
26          s_dfdy += Int((f*drdt[i,0]).subs({x:r[i,0], y:r[i,1]}),t_limits)
27
28      s_gradf = sp.simplify(sp.Matrix([s_dfdx,-s_dfdy])/A).expand()
29      s_gradf = sp.nsimplify(s_gradf, tolerance=1e-9).evalf(6)
30      if p_out:
31          display(Math(f"\\text{{Exact gradient of f at polygon center}}=\
32                          {latex(gradf)}"))
33          display(Math(f"\\text{{Smoothed gradient of f using polygon}}=\
34                          {latex(s_gradf)}"))
35      return s_gradf, gradf
```

In preparing the data for our computations, we designed two polygons. The first is a regular square, a quadrilateral with four nodes (Q4), character-ized by its geometric symmetry and four symmetric lines. The second is an irregular pentagon with five nodes (P5) crafted without geometric symmetry. The data for both polygons is as follows:

```
1  #nodes22 = np.array([[-1,-1], [ 1,-1], [ 1, 1], [-1, 1]]) # Alternative
2  nodesQ4 = np.array([[-1,-1],   # node 0, quad. with 4 nodes, Q4 (square)
3                      [ 1,-1],                              # node 1
4                      [ 1, 1],                              # node 2
5                      [-1, 1]])                             # node 3
6                                                            # A = 4
7  nodesP5 = np.array([[-1.1, -1.1],    # node 0 on Pentagon with 5 nodes
8                      [ 0.4, -1.1],               # node 1 on P5
9                      [0.85,  0.3],               # node 2 on P5
10                     [-0.1,  1.3],               # node 3 on P5
11                     [-1.6,  0.3]])              # node 4 on P5
12                                                 # A = 3.99
13 np_cntr = np.array([10, 10])           # the center of the polygon
```

To examine the effects of the size of the smoothing domain, we introduce a characteristic edge length for the polygons, defining it as a symbolic variable p. We then multiply this variable by the nodal coordinates of the polygons. This approach allows the obtained gradient to incorporate p as a parameter, reflecting its dependence on the size of the smoothing domain. The code snippet is as follows:

```
1  # smoothed gradient of the linear function:
2  x, y = symbols('x, y ', real=True)
3  p = symbols('p', real=True)     # introduce a characteristic edge length
4  X = Matrix([x, y ])
5
6  nodes = nodesP5*p + np_cntr              # using P5 as the smoothing domain
7  s_gradf, gradf = gsm_polygon(f1, X, nodes, p_out=True)
```

$$\text{Exact gradient of f at polygon center} = \begin{bmatrix} 1.0 \\ 2.0 \end{bmatrix}$$

$$\text{Smoothed gradient of f using polygon} = \begin{bmatrix} 1.0 \\ 2.0 \end{bmatrix}$$

It is observed that the gradient and the smoothed gradient of the linear function are identical regardless of the size and location of the smoothing domain. This finding is consistent with the results obtained using the circular curve.

The following code snippet computes the gradients and the smoothed gradient of the quadratic function using the `gsm_polygon()` function:

```
1  # smoothed graident of the quadratic function using P5:
2
3  s_gradf, gradf = gsm_polygon(f2, X, nodesP5*p+np_cntr, p_out=True)
4  display(Math(f"\\text{{Difference}}={latex((s_gradf-gradf).evalf(6))}"))
```

$$\text{Exact gradient of f at polygon center} = \begin{bmatrix} 221.0 - 6.82p \\ 442.0 - 2.64p \end{bmatrix}$$

$$\text{Smoothed gradient of f using polygon} = \begin{bmatrix} 221.0 - 7.45556p \\ 442.0 - 1.92982p \end{bmatrix}$$

$$\text{Difference} = \begin{bmatrix} -0.635556p \\ 0.710175p \end{bmatrix}$$

This time, we have found differences between the smoothed and exact gradients of the quadratic function. The discrepancy depends on the characteristic edge length p. If we shrink the polygon, these two gradients converge and become identical when $p = 0$. This indicates that Eq. (10.2) serves as an approximation of the exact gradient for functions of order higher than 1. We note the following:

> If the goal is to obtain a smoothed gradient that closely approximates the exact gradient of functions of order higher than 1, the size of the smoothing domain should be reduced.

This finding contrasts with our observations when using the circular domain for the quadratic function. The discrepancy arises due to the asymmetry of the smoothing domain. To confirm this, we will use the square smoothing domain Q4 instead of the irregular pentagon P5 and perform the same analysis:

```
1  # smoothed graident of the quadratic function using Q4:
2
3  s_gradf, gradf = gsm_polygon(f2, X, nodesQ4*p+np_cntr, p_out=True)
4  display(Math(f" \\text{{Difference}}={latex((s_gradf-gradf).evalf())}"))
```

$$\text{Exact gradient of f at polygon center} = \begin{bmatrix} 221.0 \\ 442.0 \end{bmatrix}$$

$$\text{Smoothed gradient of f using polygon} = \begin{bmatrix} 221.0 \\ 442.0 \end{bmatrix}$$

$$\text{Difference} = \begin{bmatrix} 0 \\ 0 \end{bmatrix}$$

We found zero difference as expected.

Let us proceed with the cubic (third-order) function:

```
1  # smoothed graident of the third-order funciton using Q4:
2
3  s_gradf, gradf = gsm_polygon(f3, X, nodesQ4*p+np_cntr, p_out=True)
4  display(Math(f" \\text{{Difference}}={latex((s_gradf-gradf).evalf())}"))
```

$$\text{Exact gradient of f at polygon center} = \begin{bmatrix} 3301.0 \\ 442.0 \end{bmatrix}$$

$$\text{Smoothed gradient of f using polygon} = \begin{bmatrix} 11.0p^2 + 3301.0 \\ 442.0 \end{bmatrix}$$

$$\text{Difference} = \begin{bmatrix} 11.0p^2 \\ 0 \end{bmatrix}$$

This time, we found a difference between the smoothed and exact gradients because the function is third order. Regardless of the type of smoothing domain used (circular, Q4, or P5), there will always be some error. If we shrink the polygon, these two gradients converge and become identical when $p = 0$. This again indicates that Eq. (10.2) serves as an approximation of the exact gradient for functions of order higher than 1.

Of course, we can reduce the difference by decreasing the size of the smoothing domain. Let us confirm this:

```
1  # smoothed graident of the 3rd-order funciton using P5, reduced 10×:
2
3  s_gradf, gradf = gsm_polygon(f3, X, nodesP5*p/10+np_cntr, p_out=True)
4  Difference = (s_gradf-gradf).simplify().evalf(6)
5  display(Math(f" \\text{{Difference}}={latex(Difference)}"))
```

$$\text{Exact gradient of f at polygon center} = \begin{bmatrix} 0.031713p^2 - 20.46p + 3301.0 \\ 442.0 - 0.264p \end{bmatrix}$$

$$\text{Smoothed gradient of f using polygon} = \begin{bmatrix} 0.142761p^2 - 22.3667p + 3301.0 \\ 442.0 - 0.192982p \end{bmatrix}$$

$$\text{Difference} = \begin{bmatrix} p\,(0.111048p - 1.90667) \\ 0.0710175p \end{bmatrix}$$

As expected, the difference is reduced even when using the P5 smoothing domain. If we further decrease the size of the polygon, the difference becomes smaller and ultimately zero when $p = 0$.

It's important to note that when using a circular smoothing domain, the types of functions that can be analytically integrated are quite limited. For example, if we were to utilize a sine or cosine function, the `gsm_circle()` function may fail to yield a solution. In such cases, the `gsm_polygon()` function should be employed instead.

Let us consider two examples: one using a large P5 polygon and another using a P5 reduced by a factor of 10, to compute the smoothed gradient of a function that combines polynomial and sinusoidal components:

```
1  f = y*sin(x) + x*cos(y)
2  s_gradf, gradf = gsm_polygon(f, X, nodesP5+np_cntr, p_out=True)
3  display(Math(f" \\text{{Difference}}={latex((s_gradf-gradf).evalf())}"))
```

$$\text{Exact gradient of f at polygon center} = \begin{bmatrix} -10.4626 \\ 4.51241 \end{bmatrix}$$

$$\text{Smoothed gradient of f using polygon} = \begin{bmatrix} -8.94207 \\ 3.93274 \end{bmatrix}$$

$$\text{Difference} = \begin{bmatrix} 1.52055168151855 \\ -0.579667568206787 \end{bmatrix}$$

We observe again some differences. Let us shrink the size of the smoothing domain:

```
1  f = y*sin(x) + x*cos(y)
2  s_gradf, gradf = gsm_polygon(f, X, nodesP5/10+np_cntr, p_out=True)
3  display(Math(f" \\text{{Difference}}={latex((s_gradf-gradf).evalf())}"))
```

$$\text{Exact gradient of f at polygon center} = \begin{bmatrix} -9.39249 \\ 4.85531 \end{bmatrix}$$

$$\text{Smoothed gradient of f using polygon} = \begin{bmatrix} -9.39281 \\ 4.86182 \end{bmatrix}$$

$$\text{Difference} = \begin{bmatrix} -0.000318527221679688 \\ 0.00651168823242188 \end{bmatrix}$$

We found that the difference is significantly reduced.

Solution to Question 6: We can now summarize the findings from this case study as follows:

1. Equation (10.2) produces the exact gradient for linear scalar functions at the center of the smoothing domain used. This holds true regardless of the shape, size, or location of the smoothing domain.

2. Equation (10.2) yields the exact gradient for certain quadratic scalar functions when the smoothing domain exhibits specific symmetry properties.
3. In general, Eq. (10.2) provides an approximate gradient for functions of order higher than one or for nonpolynomial types. The accuracy of this approximation depends on the shape and size of the smoothing domain. To improve accuracy, one should reduce the size of the smoothing domain.
4. The GSM directly utilizes function values to compute the gradient, thereby avoiding differentiation. As a result, it is very stable and robust.

Additionally, we note that Eq. (10.2) provides the theoretical foundation for the GSM. It offers an effective way to approximate the gradient (derivatives) of functions and is a valuable addition to the function approximation techniques discussed in Refs. [5, 7].

The gradient smoothing techniques have been applied to a wide range of problems. In most practical applications, polygonal smoothing domains are more commonly used. This preference arises from domain discretization, which generates a mesh, resulting in polygonal shapes of smoothing domains with an arbitrary number of edges.

10.2 Computation of the smoothed curl of vector fields

The GSM can also be extended to compute the smoothed curl of vector fields. In this context, the result will be a scalar for 2D cases. Consider a vector field $\mathbf{F} = [F_x \ F_y]^\top$. Using Green's circulation theorem, we have

$$\underbrace{\frac{1}{A} \iint_{\mathcal{D}} \left(\frac{\partial F_y}{\partial x} - \frac{\partial F_x}{\partial y} \right) dx\, dy}_{\text{smoothed curl over } \mathcal{D},\ \overline{\text{curl}\mathbf{F}}} = \frac{1}{A} \oint_{\mathcal{C}} (F_x\, dx + F_y\, dy) \qquad (10.4)$$

where A is the area of domain $\mathcal{D}$. Equation (10.4) can also be expressed as

$$\overline{\text{curl}\mathbf{F}} = \frac{1}{A} \oint_{\mathcal{C}} (F_x\, dx + F_y\, dy) = \frac{1}{A} \oint_{\mathcal{C}} \mathbf{F} d\mathbf{r} \qquad (10.5)$$

Here, $\overline{\text{curl}\mathbf{F}}$ is referred to as the smoothed curl in the (x, y) plane (with respect to the z-axis) of vector $\mathbf{F}$ over the smoothing domain $\mathcal{D}$.

It is noteworthy that the smoothed curl is computed using only the values of the component functions in the vector $\mathbf{F}$ via curve integration (not differentiation). The curve $\mathcal{C}$ in Eq. (10.5) can take any shape and be located anywhere in the domain. The only requirement for $\mathcal{C}$ is that it must be at least piecewise smooth and enclose the domain $\mathcal{D}$.

10.2.1 *Case study: Smoothed curl of vector fields*

Consider vector functions with polynomial components defined in 2D space as

$$\mathbf{s}_z = [-y \ x]^\top; \quad \text{the spin field}$$

$$\mathbf{F}_2 = \begin{bmatrix} -y^2 \\ x^2 \end{bmatrix}; \quad \text{quadratic} \tag{10.6}$$

$$\mathbf{F}_3 = \begin{bmatrix} -y^3 \\ x^3 \end{bmatrix}; \quad \text{third order}$$

The curls of these functions can be computed easily through differen-tiation. However, we aim to compute their smoothed curl without using differentiation. We will consider two different smoothing domains: a circle, as shown in Fig. 10.1(a), and a polygon, as shown in Fig. 10.1(b).

The tasks for this study are as follows:

1. Compute the exact curl of these vector functions.
2. Using the circular smoothing domain shown in Fig. 10.1(a), compute the smoothed curl of the linear function $\mathbf{s}_z(x, y)$ at the center of the circle.
3. Repeat item 2 for the quadratic vector function $\mathbf{F}_2(x, y)$.
4. Repeat item 2 for the cubic (third-order) vector function $\mathbf{F}_3(x, y)$.
5. Repeat items 2–4 using the pentagon shown in Fig. 10.1(b) as the smooth-ing domain.
6. Discuss the results obtained.

Solution to Question 1: This is straightforward as it involves only partial differentiation of the functions. We will use the Python function `gr.curl_vf()` for this purpose. The following code snippet produces the answer:

```
1   # Exact solution for the curl of the vector functions:
2
3   x, y = symbols('x, y ', real=True)
4   X = sp.Matrix([x, y])
5
6   sz = sp.Matrix([-y, x])                        # Linear, spin field
7   F2 = sp.Matrix([-y**2, x**2])                        # quaduratic
8   F3 = sp.Matrix([-y**3, x**3])                          # 3rd order
9
10  display(Math(f" \\text{{Exact curl of these fields:}}"))
11  display(Math(f" \\text{{curl}}({latex(sz)})={latex(gr.curl_vf(sz, X))};\
12          \;\; \;\; \\text{{curl}}({latex(F2)})={latex(gr.curl_vf(F2, X))};\
13          \;\; \;\; \\text{{curl}}({latex(F3)})={latex(gr.curl_vf(F3, X))}"))
```

Exact curl of these fields:

$$\text{curl}\left(\begin{bmatrix} -y \\ x \end{bmatrix}\right) = 2; \quad \text{curl}\left(\begin{bmatrix} -y^2 \\ x^2 \end{bmatrix}\right) = 2x + 2y; \quad \text{curl}\left(\begin{bmatrix} -y^3 \\ x^3 \end{bmatrix}\right) = 3x^2 + 3y^2$$

As expected, the curl of the linear vector function is a constant function, the curl of the quadratic vector function is a linear function, and the curl of the third-order vector function is a quadratic function.

To answer the other questions, we write the Python function in the following section for multiple uses in our study.

10.2.2 *Code for the smoothed curl over a circle*

The following Python function computes the smoothed curl of a given vector function over a circular domain with a radius of a. The description and tasks for the code are listed in the docstring:

```python
 1  def csm_circle(vf, X, center, a, p_out=True):
 2      '''Compute the smoothed curl of a 2D vector function vf using a
 3         circular curve with radius of r=a. It prints out
 4         1 The curl of f.
 5         2 Smoothed curl of f.
 6         3 center: the center of the circle (controls the location of it)
 7      return: s_curlf, curlf -  smoothed curl, exact curl of vf.
 8      '''
 9      θ = symbols('θ', real=True)          # polar angle as the parameter
10      x, y = X                             # unpack the Cartesian coordinates
11      A = sp.pi*a**2                       # area of the circle
12
13      dicX0 = {x:center[0], y:center[1]}
14      curlf = gr.curl_vf(vf, X).subs(dicX0)           # exact curl of vf
15
16      # Coordinate transformation, for integral over the closed circle C:
17      xC = a*cos(θ) + center[0]                # xC, yC are x,y on curve C
18      yC = a*sin(θ) + center[1]
19      dicC = {x:xC, y:yC}
20
21      fIC = vf.subs(dicC)
22      #display(Math(f"\\text{{ Integrand of the curve integral}}=\
23      #                        {latex(sp.simplify(fIC))}"))
24      dr = sp.Matrix([xC.diff(θ), yC.diff(θ)])
25
26      # Compute the smoothed curl:
27      s_curlf=(sp.integrate(fIC.dot(dr),(θ,0,2*sp.pi))/A).simplify()
28
29      if p_out:                                   # print out the results
30          display(Math(f" \\text{{Exact curl of vf = }}{latex(curlf)}"))
31          display(Math(f"\\text{{Smoothed curl of vf}}={latex(s_curlf)}"))
32
33      return s_curlf, curlf
```

Solution to Question 2: We use csm_circle() to compute the smoothed curl of the linear vector function (the spin field):

```
1  a, c = symbols('a, c', nonnegative=True)      # circle radius and center
2
3  center = Matrix([2*c, 5*c])        # place the center of the sphere here
4  s_curlf, curlf = csm_circle(sz, X, center, a, p_out=True)
```

Exact curl of vf $= 2$

Smoothed curl of vf $= 2$

It is found that the smoothed curl of the linear vector function is the same as the exact curl.

Solution to Question 3: Compute the smoothed curl of the quadratic vector function using the same Python function:

```
1  s_curlf, curlf = csm_circle(F2, X, center, a, p_out=True)
```

Exact curl of vf $= 14c$

Smoothed curl of vf $= 14c$

It is found that the smoothed curl of the quadratic vector function is again the same as the exact curl. This is true regardless of the size and location of the smoothing domain, implying that Eq. (10.5) produces the exact curl for both the linear and quadratic vector functions using a circular smoothing domain.

Solution to Question 4: Compute the smoothed curl of the third-order vector function using the same Python function:

```
1  s_curlf, curlf = csm_circle(F3, X, center, a, p_out=True)
2  display(Math(f" \\text{{Difference}}={latex((s_curlf-curlf))}"))
```

Exact curl of vf $= 87c^2$

Smoothed curl of vf $= \dfrac{3a^2}{2} + 87c^2$

Difference $= \dfrac{3a^2}{2}$

This time, we found that the smoothed curl is different from the exact one. The smoothed curl depends on the radius of the circle a (quadratically). If we shrink the circle, the difference becomes smaller and is zero if $a = 0$. This means that Eq. (10.5) is an approximation of the exact curl for a vector function with polynomial component functions of order higher than 2. We note the following:

> If the purpose is to obtain a smoothed curl that is close to the exact curl of a vector function with polynomial component functions of order higher than 2, we shall reduce the smoothing domain.

Solution to Question 5: To compute the smoothed curl of these vector functions over a pentagon domain, we write the Python function in the following section.

10.2.3 *Code for the smoothed curl over a polygon*

The following Python function computes the smoothed curl of a given vector function over a general polygonal domain that consists of only straight lines. The tasks that the code performs are listed in the doc-string of the code:

```
 1  def csm_polygon(vf, X, nodes, p_out=True):
 2      '''Computes the smoothed curl of a given scalar function f in 2D
 3      over a general polygonal domain, defined by its nodes.
 4      return: s_curlf, curlf -  smoothed curl, exact curl of vf.
 5      '''
 6      x, y = X
 7      t = sp.symbols('t')                    # parameter for line integration
 8
 9      #A = gr.polygon_area(nodes)       # area of the polygon, alternative
10      A = gr.polygon_area_green(nodes)
11      curlf = gr.curl_vf(vf, X).subs(dict(zip(X, np.mean(nodes, axis=0))))
12
13      # We roll the nodes back by 1, and use it to form the polygon edges.
14      nodes_r1 = np.roll(nodes, -1, axis=0)
15      r = nodes + t*(nodes_r1-nodes)   # parameterizaton with t for edges
16      drdt = sp.diff(sp.Matrix(r), t)     # drdt[i,0]=dxdt; drdt[i,1]=dydt
17      t_ab = (t, 0, 1)                          # limits for t (0-1 for each edge)
18
19      Int = sp.integrate; s_curlf = 0
20      for i in range(len(nodes)): # line integrals over dt along each edge
21          s_curlf += Int(vf.dot(drdt[i,:]).subs({x:r[i,0],y:r[i,1]}),t_ab)
22
23      s_curlf = sp.simplify(s_curlf/A)
24      if p_out:
25          display(Math(f"\\text{{Exact curl of vf = }}{latex(curlf)}"))
26          display(Math(f"\\text{{Smoothed curl of vf}}=\
27                                {latex(s_curlf)}"))
28      return s_curlf, curlf
```

In the following studies, we use these two polygons from the previous case study: One is a regular square Q4 and the other is an irregular pentagon P5. The data for these smoothing domains are the same as before:

```
1  # smoothed curl of the linear vector funciton:
2
3  x, y = symbols('x, y ', real=True)
4  X = Matrix([x, y ])
5
6  s_curlf, curlf = csm_polygon(sz, X, nodesP5+np_cntr, p_out=True)
```

Exact curl of vf $= 2$

Smoothed curl of vf $= 2.0$

It is found that the curl and the smoothed curl of the linear vector function are identical, which is consistent with the results obtained using the circular domain.

The following snippet computes the smoothed curl of the quadratic vector function using **gsm_polygon()**:

```
1  # smoothed curl of the quadartic funciton using P5:
2
3  s_curlf, curlf = csm_polygon(F2, X, nodesP5+np_cntr, p_out=True)
4  display(Math(f" \\text{{Difference}}={latex((s_curlf-curlf).evalf())}"))
```

Exact curl of vf $= 39.26$

Smoothed curl of vf $= 39.2345029239767$

Difference $= -0.0254970760233348$

This time, we found differences between the smoothed and exact curls of the quadratic vector function, which contrasts with the results obtained using the circular domain. These differences arise due to the symmetry of the smoothing domain. To confirm this, we use the square smoothing domain Q4 instead of P5 and perform the same analysis again:

```
1  # smoothed graident of the quadartic funciton using Q4:
2
3  s_curlf, curlf = csm_polygon(F2, X, nodesQ4+np_cntr, p_out=True)
4  display(Math(f" \\text{{Difference}}={latex((s_curlf-curlf).evalf())}"))
```

Exact curl of vf $= 40.0$

Smoothed curl of vf $= 40.0$

Difference $= 0$

We found zero difference as expected.

Let us proceed with the third-order vector function:

```
1  # smoothed curl of the cubic (third-order) vector funciton using Q4:
2
3  s_curlf, curlf = csm_polygon(F3, X, nodesQ4+np_cntr, p_out=True)
4  display(Math(f" \\text{{Difference}}={latex((s_curlf-curlf).evalf())}"))
```

Exact curl of vf $= 600.0$

Smoothed curl of vf $= 602.0$

Difference $= 2.0$

We found differences between the smoothed and exact curls because the function is third order. Regardless of the type of smoothing domain used (circular, Q4, or P5), there will be some errors. In this case, we can reduce the difference by decreasing the size of the smoothing domain. Let us confirm this:

```
1  # smoothed curl of the 3rd-order vector funciton using P5, reduced 10x:
2
3  s_curlf, curlf = csm_polygon(F3, X, nodesP5/10+np_cntr, p_out=True)
4  display(Math(f" \\text{{Difference}}={latex((s_curlf-curlf).evalf())}"))
```

Exact curl of vf $= 597.782991$

Smoothed curl of vf $= 597.726481798202$

Difference $= -0.0565092017982352$

As expected, the difference is significantly reduced even when using the P5 smoothing domain.

Note that when using a circular smoothing domain, the range of functions that can be integrated is quite limited. For more complicated functions, the **gsm_polygon()** function should be used. Let us examine two examples: one using a large P5 and another using a P5 reduced by a factor of 10 to compute the smoothed curl of a function that is a composite of polynomial and sinusoidal components:

```
1  F = sp.Matrix([y*sin(x), x*cos(y)])
2  s_curlf, curlf = csm_polygon(F, X, nodesP5+np_cntr, p_out=True)
3  display(Math(f" \\text{{Difference}}={latex((s_curlf-curlf).evalf())}")
```

Exact curl of vf $= -0.60805978869881$

Smoothed curl of vf $= -0.526658859514634$

Difference $= 0.0814009291841759$

```
1  # smoothing domain is reduced by 10 times:
2
3  s_curlf, curlf = csm_polygon(F, X, nodesP5/10+np_cntr, p_out=True)
4  display(Math(f" \\text{{Difference}}={latex((s_curlf-curlf).evalf())}"))
```

Exact curl of vf $= -0.324567854707539$

Smoothed curl of vf $= -0.325594499919807$

Difference $= -0.00102664521226764$

Solution to Question 6: We can now summarize the findings from this case study as follows:

1. Equation (10.5) produces the exact curl for linear vector functions at the center of the smoothing domain. This is true regardless of the shape, size, or location of the smoothing domain.
2. Equation (10.5) produces an approximate curl for functions beyond the linear or nonpolynomial types. The accuracy of the approximation depends on the shape and size of the smoothing domain used. To increase the accuracy, one should reduce the size of the smoothing domain.

We also note that Eq. (10.5) provides the theoretical basis for the curl smoothing method. It offers an effective approach for approximating the curl of vector functions.

In most practical applications involving domain discretization, polygonal smoothing domains are preferred.

Smoothing techniques can also be used for approximating the divergence of vector functions. It is a good exercise for readers to compute the smoothed divergence of a vector function based on Eq. (8.10) or Eq. (8.11) by modifying the `csm_circle()` and `csm_polygon()` functions.

10.3 Computation of smoothed gradient in 3D

Consider a 3D smoothing domain $\mathcal{D}$ with volume V. Let $W = \frac{1}{V}$ in the smoothing domain, which leads to $\nabla W = \mathbf{0}$. The gradient integral formula Eq. (8.24) becomes

$$
\begin{bmatrix} \frac{\partial v}{\partial x} \\ \frac{\partial v}{\partial y} \\ \frac{\partial v}{\partial z} \end{bmatrix} \equiv \underbrace{\frac{1}{V} \iiint_{\mathcal{D}} (\nabla v)\, \mathrm{d}V}_{\text{a vector containing smoothed partial derivatives}} = \frac{1}{V} \oint_{S} v\, \mathbf{n}\, \mathrm{d}S \qquad (10.7)
$$

The term in the middle of the foregoing equation is an averaged gradient ∇v or smoothed gradient over domain $\mathcal{D}$. It contains three smoothed derivatives. To compute these, we use the surface integral on the right side.

Thus, the smoothed gradients or derivatives can be written explicitly as

$$
\begin{bmatrix} \frac{\partial v}{\partial x} \\ \frac{\partial v}{\partial y} \\ \frac{\partial v}{\partial z} \end{bmatrix} = \begin{bmatrix} \frac{1}{V} \oint_{S} v n_x\, \mathrm{d}S \\ \frac{1}{V} \oint_{S} v n_y\, \mathrm{d}S \\ \frac{1}{V} \oint_{S} v n_z\, \mathrm{d}S \end{bmatrix} \qquad (10.8)
$$

where n_x, n_y, n_z are the components of the outward unit normal at point (x, y, z) on the surface $\mathcal{S}$. In general, all these are functions of the coordinates.

It is seen that the smoothed partial derivatives of function v can be computed using only the values of function v on the boundary surface. This is an extension from 2D discussed earlier. The surface $\mathcal{S}$ in Eq. (10.8) can have an arbitrary shape and size, but needs to be at least piecewise smooth.

10.3.1 *Case study: Smoothed gradients of functions*

Consider scalar functions in 3D defined as

$$f_1(x, y, z) = x + 2y + 3z + 8; \quad \text{linear}$$
$$f_2(x, y, z) = 11x^2 + 22y^2 + 33z^2 + x + 2y + 3z + 8; \quad \text{quadratic}$$
$$f_3(x, y, z) = 11x^3 + 22y^3 + 33z^3 + x + 2y + 3z + 8; \quad \text{third order}$$
$$f_{np}(x, y, z) = y \sin(x) + z \cos(y); \quad \text{nonpolynomial}$$

(10.9)

The gradients of these functions can be computed with ease through differentiation. However, we would like to compute their smoothed gradients without using differentiation. We consider two different smoothing domains: a sphere shown in Fig. 10.2(a) and a brick or box shown in Fig. 10.2(b).

The tasks for this study are as follows:

1. Compute the exact solution for the gradient of these functions.
2. Using the spherical smoothing domain shown in Fig. 10.2(a), compute the smoothed gradient of $f_1(x, y, z)$ at the center of the sphere.
3. Repeat item 2 but for function $f_2(x, y, z)$.
4. Repeat item 2 but for function $f_3(x, y, z)$.
5. Repeat items 2–4 but using the brick shown in Fig. 10.2(b) as the smoothing domain.
6. Repeat item 5 for function $f_{np}(x, y, z)$.
7. Discuss the results obtained.

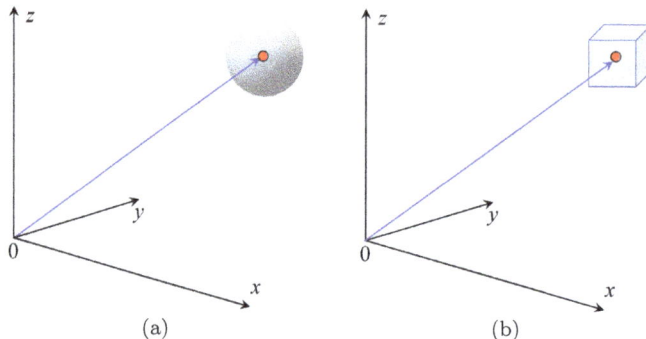

(a)   (b)

Figure 10.2. (a) A spherical and (b) a brick domain for computing the smoothed gradient of a scalar function at the center of the domains.

Solution to Question 1: This is straightforward as it involves only partial differentiation of the function. We use the following code snippet to produce the answer:

```
1  # Exact solution for the gradient of the functions:
2
3  x, y, z = symbols('x, y, z', real=True)
4  X = Matrix([x, y, z])
5
6  f1 = 1*x + 2*y + 3*z + 8                                    # linear
7  f2 = 11*x**2 + 22*y**2 + 33*z**2 + 1*x+ 2*y + 3*z + 8      # quaduratic
8  f3 = 11*x**3 + 22*y**3 + 33*z**3 + 1*x+ 2*y + 3*z + 8      # 3rd order
9  fnp = y*sin(x) + z*cos(y)                                   # nonpolynomial
10
11 display(Math(f" \\text{{Exact gradient of these functions:}}"))
12 display(Math(f" \\text{{∇}}({latex(f1)}) = {latex(gr.grad_f(f1, X))}"))
13 display(Math(f" \\text{{∇}}({latex(f2)}) = {latex(gr.grad_f(f2, X))}"))
14 display(Math(f" \\text{{∇}}({latex(f3)}) = {latex(gr.grad_f(f3, X))}"))
15 display(Math(f" \\text{{∇}}({latex(fnp)})= {latex(gr.grad_f(f3, X))}"))
```

Exact gradient of these functions:

$$\nabla(x + 2y + 3z + 8) = \begin{bmatrix} 1 \\ 2 \\ 3 \end{bmatrix}$$

$$\nabla(11x^2 + x + 22y^2 + 2y + 33z^2 + 3z + 8) = \begin{bmatrix} 22x + 1 \\ 44y + 2 \\ 66z + 3 \end{bmatrix}$$

$$\nabla(11x^3 + x + 22y^3 + 2y + 33z^3 + 3z + 8) = \begin{bmatrix} 33x^2 + 1 \\ 66y^2 + 2 \\ 99z^2 + 3 \end{bmatrix}$$

$$\nabla(y \sin(x) + z \cos(y)) = \begin{bmatrix} 33x^2 + 1 \\ 66y^2 + 2 \\ 99z^2 + 3 \end{bmatrix}$$

To answer the remaining questions, we write the Python function in the following section.

10.3.2 *Code for the smoothed gradient over a sphere*

The following Python function computes the smoothed gradient of a given scalar function $f(x, y, z)$ over a spherical domain with radius a. The tasks for the code are listed in the docstring:

```
 1  def gsm_sphere(f, X, a, center, p_int=True):
 2      '''Compute the smoothed gradient of a scalar function f using a
 3          spherical surface with radius a. It prints out:
 4          1 The gradient of f.
 5          2 Smoothed gradient of f.
 6          center: the center of the sphere.
 7          p_int: controls print out.
 8          return: s_gradf - smoothed gradient of
 9                  gradf - gradient of f
10      '''
11      r = sp.symbols('r', nonnegative=True)  # radial coordinate, spheres
12      θ, φ = symbols('θ, φ', real=True)
13      x, y, z = X
14
15      # Integral over an enclosed sphere:
16      x_ = r*sin(θ)*cos(φ) # x_, y_, and z_ are x,y,z on sphere surface
17      y_ = r*sin(θ)*sin(φ)
18      z_ = r*cos(θ)
19      dic = {x:x_, y:y_, z:z_}
20
21      #V_ball = 4*sp.pi*a**3/3 # volume of sphere of a; used for checking
22      # Volume integration over the spherical ball with r=a:
23      fID = r**2 *sin(θ)         # unit function for computing the volume
24      Int = sp.integrate; pi = sp.pi
25      V_ball = Int(Int(Int(fID,(r,0,a)),(θ,0,pi)),(φ,0,2*pi)).simplify()
26
27      dicX0 = {x:center[0], y:center[1], z:center[2]}
28      gradf = gr.grad_f(f, X).subs(dicX0)
29
30      n = X.subs(dic)/a         # normal n on S: n = 1/r on sphere surface
31      dicX = {x:x_+center[0], y:y_+center[1], z:z_+center[2]}
32      f_shift = f.subs(dicX)
33      fSI = sp.simplify((f_shift*n*r**2*sin(θ)).subs(dic).subs(r,a))
34
35      s_gradf = (Int(Int(fSI,(φ,0,2*pi)),(θ,0,pi))/V_ball).simplify()
36
37      if p_int:                                    # print out the results
38          display(Math(f" \\text{{Gradient of f = }}{latex(gradf.T)}"))
39          display(Math(f" \\text{{Smoothed gradient of f}}=\
40                          {latex(s_gradf.T)}"))
41      return s_gradf, gradf
```

Solution to Question 2: Compute the gradient and the smoothed gradient of the linear function using the Python function **gsm_sphere**():

```
1  x, y, z = symbols('x, y, z', real=True)
2  a = symbols('a', nonnegative=True)            # radius of the sphere
3  X = Matrix([x, y, z])
4
5  center = Matrix([2*a, 5*a, 3*a])  # place the center of the sphere here
6  s_gradf, gradf = gsm_sphere(f1, X, a, center, p_int=True)
```

Gradient of f $= \begin{bmatrix} 1 & 2 & 3 \end{bmatrix}$

Smoothed gradient of f $= \begin{bmatrix} 1 & 2 & 3 \end{bmatrix}$

It is found that the smoothed gradient of the linear function is the same as its exact gradient. This is true regardless of the radius and location of the smoothing domain.

Solution to Question 3: Compute the smoothed gradient of the quadratic function using the same Python function:

```
1  # smoothed gradient of the quadratic function:
2  s_gradf, gradf = gsm_sphere(f2, X, a, center, p_int=True)
```

Gradient of f $= \begin{bmatrix} 44a + 1 & 220a + 2 & 198a + 3 \end{bmatrix}$

Smoothed gradient of f $= \begin{bmatrix} 44a + 1 & 220a + 2 & 198a + 3 \end{bmatrix}$

In the results found above, c controls the location of the sphere, while a controls its size. It is found that the smoothed gradient of the quadratic function is the same as the exact gradient at the center of the smoothing domain. This holds true regardless of the size or location of the smoothing domain. Thus, Eq. (10.8) produces the exact gradient for both the linear and quadratic scalar functions.

Solution to Question 4: Compute the smoothed gradient of the third-order function using the same Python function:

```
1  # smoothed graident of the third-order funciton:
2  s_gradf, gradf = gsm_sphere(f3, X, a, center, p_int=True)
3  display(Math(f" \\text{{Difference}}=\
4                      {latex((s_gradf-gradf).evalf().T)}"))
```

Gradient of f $= \begin{bmatrix} 132a^2 + 1 & 1650a^2 + 2 & 891a^2 + 3 \end{bmatrix}$

Smoothed gradient of f $= \begin{bmatrix} \frac{693a^2}{5} + 1 & \frac{8316a^2}{5} + 2 & \frac{4554a^2}{5} + 3 \end{bmatrix}$

Difference $= \begin{bmatrix} 6.6a^2 & 13.2a^2 & 19.8a^2 \end{bmatrix}$

This time, we found that the smoothed gradient differs from the exact one. The difference depends quadratically on the radius of the circle a. If we reduce the size of the circle, the difference becomes smaller, reaching zero when $a = 0$. This indicates that Eq. (10.8) serves as an approximation of the exact gradient for functions of order higher than 2. Therefore, if the goal is to obtain a smoothed gradient close to the exact gradient of polynomial functions with an order greater than 2, we should reduce the size of the smoothing domain.

Solution to Question 5: To compute the smoothed gradient of these functions over a brick domain, we write the Python function in the following section.

10.3.3 *Code for the smoothed gradient over a brick*

The following Python function computes the smoothed gradient of a given scalar function $f(x, y, z)$ over a brick domain. The tasks that the code performs are listed in the docstring of the code. It will be used multiple times in our study:

```python
 1 def gsm_brick(f, X, node_000, node_abc, p_int=True):
 2     '''Compute the smoothed gradient of a scalar function f using the
 3        six surfaces of a brick defined by:
 4        node_000: the node at back, left, bottom of the brick
 5        node_abc: the node at front, right, top of the brick
 6     It prints out:
 7        1 The gradient of f at the center of the brick.
 8        2 Smoothed gradient of f.
 9        p_int: controls printing outs.
10        return: s_gradf - smoothed gradient of f.
11                gradf - gradient of f.
12     '''
13     x, y, z = X
14     x1, y1, z1 = node_000                      # node at back, left, bottom
15     x2, y2, z2 = node_abc                      # node at front, right, top
16
17     gradf = gr.grad_f(f, X).subs({x:(x1+x2)/2,y:(y1+y2)/2,z:(z1+z2)/2})
18
19     V_brick = (x2-x1)*(y2-y1)*(z2-z1)                     # Brick volume
20
21     Int = sp.integrate                              # Face at x = x1
22     I1 = Int(Int(f.subs(x,x1)*sp.Matrix([-1,0,0]), (y,y1,y2)),(z,z1,z2))
23     # Face at x = x2
24     I2 = Int(Int(f.subs(x,x2)*sp.Matrix([1, 0,0]), (y,y1,y2)),(z,z1,z2))
25     # Face at y = y1
26     I3 = Int(Int(f.subs(y,y1)*sp.Matrix([0,-1,0]), (x,x1,x2)),(z,z1,z2))
27     # Face at y = y2
28     I4 = Int(Int(f.subs(y,y2)*sp.Matrix([0, 1,0]), (x,x1,x2)),(z,z1,z2))
29     # Face at z = z1
30     I5 = Int(Int(f.subs(z,z1)*sp.Matrix([0,0,-1]), (x,x1,x2)),(y,y1,y2))
31     # Face at z = z2
32     I6 = Int(Int(f.subs(z,z2)*sp.Matrix([0,0, 1]), (x,x1,x2)),(y,y1,y2))
33
34     # Sum all Is to get the total surface integral:
35     s_gradf = ((I1 + I2 + I3 + I4 + I5 + I6)/V_brick).simplify()
36
37     if p_int:                                       # print out the results
38         display(Math(f" \\text{{Gradient of f = }}{latex(gradf.T)}"))
39         display(Math(f" \\text{{Smoothed gradient of f}}=\
40                     {latex(s_gradf.T)}"))
41     return s_gradf, gradf
```

Solution to Question 6: The following snippet computes the gradient and the smoothed gradient of the linear function using the Python function gsm_brick():

```
1  # smoothed graident of the linear funciton:
2  a, b, c = symbols('a, b, c', nonnegative=True) # Control the brick size
3
4  node_0 = np.array([-a,-b,-c])              # node at back, left, bottom
5  node_1 = np.array([ a, b, c])              # node at front, right, top
6  cntr = np.array([10, 10, 10])     # brick center, controls the location
7
8  s_gradf, gradf = gsm_brick(f1, X, node_0+cntr, node_1+cntr, p_int=True)
```

Gradient of f $= \begin{bmatrix} 1 & 2 & 3 \end{bmatrix}$

Smoothed gradient of f $= \begin{bmatrix} 1 & 2 & 3 \end{bmatrix}$

It is found that the gradient and the smoothed gradient of the linear function are the same regardless of the size and location of the smoothing domain. This is consistent with the earlier case using spherical domains.

The following snippet computes the smoothed gradient of the quadratic function using **gsm_brick()**:

```
1  # smoothed graident of the quadratic funciton:
2  s_gradf, gradf = gsm_brick(f2, X, node_0+cntr, node_1+cntr, p_int=True)
```

Gradient of f $= \begin{bmatrix} 221 & 442 & 663 \end{bmatrix}$

Smoothed gradient of f $= \begin{bmatrix} 221 & 442 & 663 \end{bmatrix}$

The results are still the same. This is due to the symmetric smoothing domain used.

Let us take a look at the case of the third-order function:

```
1  # smoothed graident of the third-order funciton:
2  s_gradf, gradf = gsm_brick(f3, X, node_0+cntr, node_1+cntr, p_int=True)
3  display(Math(f" \\text{{Difference}}={latex((s_gradf-gradf).T)}"))
```

Gradient of f $= \begin{bmatrix} 3301 & 6602 & 9903 \end{bmatrix}$

Smoothed gradient of f $= \begin{bmatrix} 11a^2 + 3301 & 22b^2 + 6602 & 33c^2 + 9903 \end{bmatrix}$

Difference $= \begin{bmatrix} 11a^2 & 22b^2 & 33c^2 \end{bmatrix}$

This time, there are differences. However, these differences depend on the size of the brick (also quadratically). If we reduce the brick size, the differences decrease. If $a = b = c = 0$, the difference will be zero. Therefore, in general, the accuracy of the smoothed gradient depends on the size of the smoothing domain used.

When using a spherical smoothing domain, the types of functions that can be integrated are limited (unless we resort to numerical integration). For more complex functions, the gsm_brick() function should be used.

Let us see an example of a composite function made of polynomial and sinusoidal terms:

```
1  fnp = y*sin(x) + z*cos(y)                              # nonpolynomial
2
3  s_gradf, gradf = gsm_brick(fnp, X, node_0+cntr, node_1+cntr, p_int=True)
4  display(Math(f" \\text{{Difference}}={latex((s_gradf-gradf))}"))
```

Gradient of f $= \begin{bmatrix} 10\cos{(10)} & -9\sin{(10)} & \cos{(10)} \end{bmatrix}$

Smoothed gradient of f

$$= \begin{bmatrix} \frac{10\sin{(a)}\cos{(10)}}{a} & \frac{(-10a\sin{(b)}+b\sin{(a)})\sin{(10)}}{ab} & \frac{\sin{(b)}\cos{(10)}}{b} \end{bmatrix}$$

$$\text{Difference} = \begin{bmatrix} -10\cos{(10)} + \frac{10\sin{(a)}\cos{(10)}}{a} \\ 9\sin{(10)} + \frac{(-10a\sin{(b)}+b\sin{(a)})\sin{(10)}}{ab} \\ -\cos{(10)} + \frac{\sin{(b)}\cos{(10)}}{b} \end{bmatrix}$$

In this case, the results are quite complicated, but we can check if they are correct as follows. First, examine the smoothed gradient as the brick size approaches zero:

```
1  [sp.limit(sgf.subs({b:a, c:a}), a, 0) for sgf in s_gradf]
```

$[10\cos{(10)}, \; -9\sin{(10)}, \; \cos{(10)}]$

It is observed that it matches the exact gradient in the limit. We can also compute the differences using a large brick:

```
1  (s_gradf-gradf).subs({a:1, b:1, c:1}).evalf()
```

$$\begin{bmatrix} 1.33017183180222 \\ -0.776188178577053 \\ 0.133017183180222 \end{bmatrix}$$

The difference is quite large as expected. Let us reduce the size of the brick:

```
1  (s_gradf-gradf).subs({a:0.01, b:0.01, c:0.01}).evalf()
```

$$\begin{bmatrix} 0.000139844555621405 \\ -8.16027586187979 \cdot 10^{-5} \\ 1.39844555621591 \cdot 10^{-5} \end{bmatrix}$$

The differences are drastically reduced.

Solution to Question 7: We now summarize the findings from this case study as follows:

1. Equation (10.8) produces the exact gradient for linear scalar functions at the center of the smoothing domain. This is true regardless of the shape, size, and location of the smoothing domain.
2. Equation (10.8) produces an approximate gradient for polynomial functions beyond linear or for nonpolynomial functions. The accuracy of the approximation depends on the shape and size of the smoothing domain. To increase the accuracy, one should reduce the size of the smoothing domain. When the domain size approaches zero, the smoothed gradient approaches the exact gradient.
3. The GSM directly uses the function values to compute the gradient, avoiding differentiation. Hence, it is very stable and robust for irregular meshes.

We also mention that Eq. (10.8) provides the theoretical basis for the GSM in 3D. It offers an effective method for approximating the gradient (derivatives) of functions.

10.4 Concluding remarks

This chapter studies the GSM, which is based on Gauss's formula. We close the chapter with the following remarks:

- The GSM produces the exact partial derivatives of a function without differentiation using a boundary integral and then letting the domain size approach zero.
- Gradient and curl smoothing techniques with Python codes are presented in detail using different types of smoothing domains.
- GSM is a valuable tool for approximating the partial derivatives of a function within a small smoothing domain without resorting to differentiation.

It is an important addition to the function approximations discussed in Refs. [5, 7].

- The gradient smoothing techniques have been applied to a wide range of problems in 3D domains, including heat transfer [1, 2], acoustics [1, 2], fluid flows [8], and solid mechanics [1–3]. In most practical applications, polyhedral smoothing domains are more commonly used.

With the GSM, one can easily approximate higher-order partial derivatives. We simply need to treat the lower-order partial derivatives as the given functions.

Now that we have powerful tools for approximating unknown functions (discussed in the previous chapter and in Refs. [5, 7]) and powerful tools for approximating the derivatives of approximated functions discussed in this chapter, when combined with Gauss's formulas and their extensions discussed in the previous chapter, we have a strong framework for building modern computational methods, such as the S-FEMs [2] and the S-PIMs [9].

References

[1]   G.R. Liu, *Mesh Free Methods: Moving Beyond the Finite Element Method*, Taylor and Francis Group, New York, 2010.

[2]   G.R. Liu and T.T. Nguyen, *Smoothed Finite Element Methods*, Taylor and Francis Group, New York, 2010.

[3]   G.R. Liu and Y. Gu, *An Introduction to Meshfree Methods and Their Programming*, The Netherlands, 2005.

[4]   G. Strang, *Calculus*, Wellesley-Cambridge Press, Massachusetts, 1991.

[5]   G.R. Liu, *Numbers and Functions: Theory, Formulation, and Python Codes*, World Scientific, New Jersey, 2024.

[6]   G.R. Liu, *Mechanics of Materials: Formulations and Solutions with Python*, World Scientific, New Jersey, 2024.

[7]   G.R. Liu, *Calculus: A Practical Course with Python*, World Scientific, New Jersey, 2025.

[8]   G.R. Liu and Zirui Mao, *Gradient Smoothing Methods with Programming: Applications to Fluids and Landslides*, World Scientific, New Jersey, 2023.

[9]   G.R. Liu and Gui-Yong, Zhang, *Smoothed point interpolation methods: G space theory and weakened weak forms*, World Scientific, New Jersey, 2013.

Index

www.ingramcontent.com/pod-product-compliance
Lightning Source LLC
Chambersburg PA
CBHW081512190326
41458CB00015B/5349